T0192273

Universitext

For other titles in this series, go to
www.springer.com/series/223

Alexandre V. Borovik • Anna Borovik

Mirrors and Reflections

The Geometry of Finite Reflection Groups

 Springer

Alexandre V. Borovik
School of Mathematics
University of Manchester
Manchester, M13 9PL
United Kingdom
borovik@manchester.ac.uk
www.borovik.net/alexandre

Anna Borovik
anna@borovik.net
www.borovik.net/anna

ISBN 978-0-387-79065-7   e-ISBN 978-0-387-79066-4
DOI 10.1007/978-0-387-79066-4
Springer New York Dordrecht Heidelberg London

Library of Congress Control Number: 2009939112

Mathematics Subject Classification (2000): 20F55, 22E40, 51F15

Springer is part of Springer Science+Business Media (www.springer.com)

Preface

This expository text contains an elementary treatment of finite groups generated by reflections. There are many good books on this subject; in particular, a book by Humphreys [Hum] provides an excellent introduction to the theory, which is very much alive and under active development; see, for example, the recent survey by Dolgachev [Dol].

The main reason why we decided to write another text is not mathematical but pedagogical: we wished to emphasize the intuitive elementary geometric aspects of the theory of reflection groups. In the theory of reflection groups, the underlying ideas of many proofs can be presented by simple drawings much better than by a dry verbal exposition. Probably for the reason of their extreme simplicity these elementary arguments are mentioned in most books only briefly and tangentially.

The second reason for the existence of this book is a remarkable feature of the theory of reflection groups: its principal objects can be defined right on the spot in the most intuitive way. We give first an informal description:

> Imagine a few semitransparent mirrors in ordinary three-dimensional space. Mirrors (more precisely, their images) multiply by reflecting in each other, as in a kaleidoscope or a hall of mirrors. Of special interest are systems of mirrors that generate only finitely many reflected images. Such finite systems of mirrors happen to be one of the cornerstones of modern mathematics and lie at the core of many mathematical theories.

As usual, the theory is actually concerned with the more general case of n-dimensional Euclidean space, with mirrors being $(n - 1)$-dimensional hyperplanes rather than two-dimensional planes. To that end, we give a formal definition:

> A system of hyperplanes (mirrors) Σ in Euclidean space $\mathbb{R}^n$ is called *closed* if for any two mirrors M_1 and M_2 in Σ, the mirror image of M_2 in M_1 also belongs to Σ.

Thus, the principal objects of the theory are *finite closed systems of mirrors*. In more general terms, the theory can be described as the geometry of multiple mirror images.

Instead of a closed mirror system, one can operate with the group of transformations generated by all its reflections—and this is a more traditional approach to the theory. We prefer to emphasize the role of mirrors, since we hope that this allows us to fully engage the reader's geometric intuition. This approach is well known and exploited in Chapter 5, §3 of Bourbaki's classical text [Bou]; see also Vinberg's exposition [Vin]. We have combined it with Tits's theory of chamber complexes [Tits] and thus made the exposition of the theory almost entirely geometrical.

Finally, we cannot escape the fact that the theory of finite reflection groups leads to their full classification. In the resulting list, two of the four infinite series of reflections groups, the symmetric groups $\mathrm{Sym}_{n+1} = A_n$ and hyperoctahedral groups BC_n, are groups of symmetries of two of the most common regular polytopes, the regular n-dimensional simplex and the n-dimensional cube. The third series, D_n, is a slight modification of BC_n, while the fourth one, $G_2(n)$, is the group of symmetries of the regular planar n-gon. Therefore the theory of reflection groups mostly deals with very concrete objects; why should we avoid an equally concrete down-to-earth approach to its development?

This is why we tried to include (frequently in the form of exercises) as many elementary facts about concrete groups as possible. We feel that this is well justified even if judged from the "global" viewpoint of mathematics as a whole. Indeed, finite reflection groups form one of the cornerstones of modern algebra and geometry. Even the simplest observations about particular groups have fundamental implications, for example for the structure of Lie groups and representation theory. A mathematician working in one such area normally maintains a whole menagerie of facts about reflection groups. We believe that students will find them interesting and amusing.

We hope that our approach allows the novice mathematician easy access to the theory of reflection groups. This aspect of the book makes it close to Grove and Benson [GB]. We realize, however, that, since classical geometry has almost completely disappeared from schools and university curricula, we need to smuggle it back in and provide the student reader with a modicum of Euclidean geometry and the theory of convex polyhedra. We do not wish to appeal to the reader's geometric intuition without trying first to help him or her develop it.

In particular, we decided to saturate the book with visual material. Our sketches and diagrams are intentionally left very unsophisticated; the book was tested in a lecture course at the University of Manchester, and most pictures, in their even less sophisticated versions, were first drawn on the blackboard. There was no point in drawing pictures that could not be reproduced by students and reused in their homework. Pictures are not for decoration, they are indispensable (though perhaps greasy and soiled) tools of the trade.

This is our conscious choice; indeed, what matters—and it is part of our teaching philosophy—is that the pictures are *reproducible*. Even if the reader has very modest drawing skills, he or she should be able to draw similar pictures as a way of facilitating his or her mathematical work. Moreover, we even included in our book a short chapter "The Forgotten Art of Blackboard Drawing." Without attempting to reinvent descriptive geometry, we give there some advice on making usable mathematical drawings;

the crucial piece of advice is that you have to treat your sketch as a mathematical object.

This book contains a number of exercises of different levels of difficulty; a star * marks more difficult exercises. Some of the exercises may look irrelevant to the subject of the book and are included for the sole purpose of developing the student's geometric intuition.

Outline of the Book and Dependencies between Chapters

Part I contains mostly standard material from linear algebra, with a brief discussion of polyhedra and polyhedral cones, topics usually addressed in courses on linear programming (Chapters 3 and 4). A more experienced reader can skip this material and later return to it for reference.

The book as such starts in Part II. In Chapter 5, we introduce reflections and their mirrors, and in Chapter 6, the main concept of the book: closed systems of mirrors. Chapter 7 contains some very elementary group theory, namely discussion of the structure of dihedral groups. Geometrically, they are finite systems of mirrors in the Euclidean plane. Chapter 8 introduces dual objects, namely root systems. Following our principles to put as much flesh on the theoretical bones as possible, Chapter 9 discusses, in great but elementary detail, the root and mirror systems A_{n-1}, BC_n, D_n.

Mirrors cut a space into *chambers*; the set of all such chambers, together with the action of the reflection group on it, is known as a *Coxeter complex*. Part III of the book studies Coxeter complexes. Chapter 14 deviates somewhat from the classical treatment of reflection groups and can be skipped on first reading.

Part IV deals with the classification of reflection groups; Chapter 17 contains lists of root systems and their detailed properties.

The novice reader may wish first to build some geometric intuition; in that case, we advise him or her to read first Chapters 6 and 7 and then move to Part V. The latter contains an independent and elementary treatment of 3-dimensional reflection groups. To read it, one needs only some basic linear algebra and group theory, and the more technical material from Part I can be temporarily skipped.

Prerequisites and Use as a Course Text

This book was carefully designed to be accessible to graduate and senior undergraduate students; we tried to calibrate its level to be usable as a course text in third and fourth year of master of mathematics degree courses in England. Hence the formal prerequisites for reading the book are very modest.

We assume the reader's solid knowledge of linear algebra, especially the theory of orthogonal transformations in real Euclidean spaces.

We use a modicum of topological concepts: open and closed subsets in the Euclidean space $\mathbb{R}^n$.

We also assume that the reader is familiar with the following basic notions of group theory:

the order of a finite group; normal subgroups and factor groups; homomorphisms and isomorphisms; generators and relations; standard notation for permutations and rules for their multiplication; cyclic groups; action of a group on a set; the orbit-stabilizer theorem.

You can find this material in any introductory text on the subject. We highly recommend a book by Armstrong [Arm] for a first reading.

If this book is used as a text for a lecture course, then Grove and Benson [GB] and Humphreys [Hum] are obvious sources for auxiliary reading and extension of topics touched on in this book. In a few cases, our book borrows mathematical ideas from the other two, although our pedagogical approach is, as a rule, different.

Acknowledgments

The early versions of the text were carefully read by Robert Sandling, Richard Booth, Neil White, and Alexander Elashvili, who suggested many corrections and improvements.

The first author expresses his special thanks to his PhD students Christine Altseimer, Ayşe Berkman, and Richard Booth. Although working on completely different projects in combinatorics and group theory, each of them spent many hours toiling at the diagram of the root system of type B_3 (Figure 9.6 on page 71), which was drawn (and stayed there for months) at the blackboard in his office.

Part of the material from this book was used in the monograph [BGW]; we thank Israel Gelfand and Neil White whose comments helped to improve the present book, too.

In July 2007, the book was tested in an informal intensive course for Turkish students in Nesin Matematik Köyü in the village of Şirince in Turkey, and in April 2009 – in lectures to undegraduate students of MidEastern Technical University in Ankara arranged under provisions of the Erasmus Teaching Mobility Programme.

Our special thanks go to David Kramer for his very careful editing of the text.

Manchester,   *Alexandre Borovik*
22 June 2009   *Anna Borovik*

Contents

Part III Coxeter Complexes

Part IV Classification

Part V Three-Dimensional Reflection Groups

Part VI Appendices

Part I

Geometric Background

1

Affine Euclidean Space $A\mathbb{R}^n$

How to Read this Chapter

Since we develop a geometric approach to reflection groups, we have to use some geometry.

This chapter provides only a very sketchy description of affine geometry and can be skipped if the reader is familiar with this standard topic of linear algebra; otherwise, it would make a good exercise to reconstruct the proofs that are only outlined in our text. Notice that the chapter contains nothing new in comparison with most standard courses on linear algebra. Also, if the reader benefited from a traditional course in analytic geometry, he or she should find the material very familiar: we simply transfer to n dimensions familiar concepts of three-dimensional geometry.

The reader who wishes to understand the rest of the book can rely on his or her three-dimensional geometric intuition. The theory of reflection groups and associated geometric objects, root systems, has the most fortunate property that almost all computations and considerations can be reduced to two- and three-dimensional configurations. We shall make every effort to emphasize this intuitive geometric aspect of the theory. But as a warning to the novice reader, we wish to remind you that your intuition will bear fruit only when supported by your ability to prove rigorously "intuitively evident" facts.

For example, when you use the fact that

an $(n-1)$-dimensional linear subspace H of the n-dimensional vector space V over the real numbers $\mathbb{R}$ divides V into two open half-spaces,

it usually helps to visualize the situation by thinking about a plane in the ordinary three-dimensional real vector space $\mathbb{R}^3$, or, simplifying the situation further, about a line in a plane, which obviously divides the space (correspondingly, plane) into two halves.

A.V. Borovik and A. Borovik, *Mirrors and Reflections: The Geometry of Finite Reflection Groups*, Universitext, DOI 10.1007/978-0-387-79066-4_1,

1.1 Euclidean Space $\mathbb{R}^n$

Let $\mathbb{R}^n$ be the Euclidean n-dimensional real vector space with canonical scalar product
"$\cdot$". We identify $\mathbb{R}^n$ with the set of all column vectors

$$\alpha = \begin{pmatrix} a_1 \\ \vdots \\ a_n \end{pmatrix}$$

of length n over $\mathbb{R}$, with componentwise addition and multiplication by scalars, and
the scalar product

$$\alpha \cdot \beta = \alpha^t \beta = (a_1, \ldots, a_n) \begin{pmatrix} b_1 \\ \vdots \\ a_n \end{pmatrix} = a_1 b_1 + \cdots + a_n b_n;$$

here ^{t} denotes taking the transposed matrix. It looks a bit awkward that we arrange
the coordinates of points in rows, and the coordinates of vectors in columns. The row
notation is more convenient typographically, but since we use left notation for group
actions, we have to use column vectors: if A is a square matrix and α a vector, the
notation $A\alpha$ for the product of A and α requires α to be a column vector.

This means that we fix the canonical orthonormal basis $\epsilon_1, \ldots, \epsilon_n$ in $\mathbb{R}^n$, where

$$\epsilon_i = \begin{pmatrix} 0 \\ \vdots \\ 1 \\ \vdots \\ 0 \end{pmatrix} \quad \text{(the entry 1 is in the ith row)} .$$

The *length* $|\alpha|$ of a vector α is defined as $|\alpha| = \sqrt{a \cdot a}$. The *angle* C between two
vectors α and β is defined by the formula

$$\cos C = \frac{\alpha \cdot \beta}{|\alpha||\beta|}, \quad 0 \leqslant C \leqslant \pi.$$

The definition of C makes sense because of the Cauchy-Schwarz inequality:

$$|\alpha \cdot \beta| \leqslant |\alpha||\beta|;$$

it ensures that the right-hand side of the equation above lies in the interval $[-1, 1]$.
If $\alpha \in \mathbb{R}^n$, then
$$\alpha^{\perp} = \{ \beta \in \mathbb{R}^n \mid \alpha \cdot \beta = 0 \}$$

is the linear subspace *normal* (or *perpendicular*, or *orthogonal*) to α. If $\alpha \neq 0$ then
$\dim \alpha^{\perp} = n - 1$.

1.2 Affine Euclidean Space $\mathbb{AR}^n$

The real affine Euclidean space $\mathbb{AR}^n$ is simply the set of all n-tuples $a_1, \ldots, a_n$ of real numbers; we call them *points*.

With any two points a and b we can associate a vector in $\mathbb{R}^n$

$$\vec{ab} = \begin{pmatrix} b_1 - a_1 \\ \vdots \\ b_n - a_n \end{pmatrix}.$$

If a is a point and α a vector, $a + \alpha$ denotes the unique point b such that $\vec{ab} = \alpha$. The point a will be called the *initial point*, and b the *terminal point* of the vector $\vec{ab}$.

The real Euclidean space $\mathbb{R}^n$ models what physicists call the system of *free vectors*, i.e., physical quantities characterized by their magnitude and direction, but whose application point is of no consequence. The *n-dimensional affine Euclidean space* $\mathbb{AR}^n$ is a mathematical model of the system of *bound* vectors, that is, vectors having fixed points of application. For us, it will sometimes be convenient to make the distinction between points and vectors; as a rule, we denote points by lowercase italic letters $a, b, \ldots, y, z$, while we use lowercase Greek letters $\alpha, \beta, \ldots$ for vectors.

If $a = (a_1, \ldots, a_n)$ and $b = (b_1, \ldots, b_n)$ are two points, the *distance* $d(a, b)$ between them is defined by the formula

$$d(a, b) = \sqrt{(a_1 - b_1)^2 + \cdots + (a_n - b_n)^2}.$$

Notice that

$$d(a, b) = |\vec{ab}|.$$

One of the most basic and standard facts in mathematics is that this distance satisfies the usual axioms for a metric: for all $a, b, c \in \mathbb{AR}^n$,

- $d(a, b) \geqslant 0$;
- $d(a, b) = 0$ if and only if $a = b$;
- $d(a, b) + d(b, c) \geqslant d(a, c)$ (the triangle inequality).

Hence $\mathbb{AR}^n$ is a metric space; from time to time we shall use various topological properties of $\mathbb{AR}^n$ as a metric space. For example, if X is a subset of $\mathbb{AR}^n$, a point $x \in X$ is said to be *interior* in X if some small *ball* centered at x,

$$B(x, \delta) = \{ y \in \mathbb{AR}^n \mid d(x, y) < \delta \}$$

belongs to X. The set of all interior points of X is denoted by X° and is called the *interior* of X. The set X is *open* if $X = X^\circ$, and *closed* if its complement in $\mathbb{AR}^n$ open. It is a standard fact of topology that the intersection of a family of closed subsets is closed. In particular, there exists a unique smallest closed subset $\bar{X}$ containing X; it is called the *closure* of X. Finally, the set

$$\partial X = \bar{X} \smallsetminus X^\circ$$

is called the *boundary* of X, and points in ∂X are called *boundary points* of X.

1.3 Affine Subspaces

1.3.1 Subspaces

If U is a vector subspace in $\mathbb{R}^n$ and a is a point in $\mathbb{AR}^n$ then the set

$$a + U = \{ a + \beta \mid \beta \in U \}$$

is called an *affine subspace* in $\mathbb{AR}^n$. The *dimension* $\dim A$ of the affine subspace $A = a + U$ is the dimension of the vector space U. The *codimension* of an affine subspace A is $n - \dim A$.

If A is an affine subspace and $a \in A$ a point, then the set of vectors

$$\overrightarrow{A} = \{ \overrightarrow{ab} \mid b \in A \}$$

is a vector subspace in $\mathbb{R}^n$; it coincides with the set

$$\{ \overrightarrow{bc} \mid b, c \in A \}$$

and thus does not depend on the choice of the point $a \in A$. We shall call $\overrightarrow{A}$ the *vector space* of A. Notice that $A = a + \overrightarrow{A}$ for any point $a \in A$. Two affine subspaces A and B of the same dimension are *parallel* if $\overrightarrow{A} = \overrightarrow{B}$.

1.3.2 Systems of Linear Equations

The standard theory of systems of simultaneous linear equations characterizes affine subspaces as solution sets of systems of linear equations

$$a_{11}x_1 + \cdots + a_{1n}x_n = c_1,$$
$$a_{21}x_1 + \cdots + a_{2n}x_n = c_2,$$
$$\vdots \quad \vdots$$
$$a_{m1}x_1 + \cdots + a_{mn}x_n = c_m.$$

In particular, the intersection of affine subspaces is either an affine subspace or the empty set. The *codimension* of the subspace given by the system of linear equations is the maximal number of linearly independent equations in the system.

1.3.3 Points and Lines

Points in $\mathbb{AR}^n$ are 0-dimensional affine subspaces.

Affine subspaces of dimension 1 are called *straight lines* or *lines*. They have the form

$$a + \mathbb{R}\alpha = \{ a + t\alpha \mid t \in \mathbb{R} \},$$

where a is a point and α a nonzero vector. For any two distinct points $a, b \in \mathbb{AR}^n$ there is a unique line passing through them, that is, $a + \mathbb{R}\overrightarrow{ab}$. The *segment* $[a, b]$ is the set

$$[a, b] = \{ a + t\overrightarrow{ab} \mid 0 \leqslant t \leqslant 1 \};$$

the *open interval* (a, b) is the set

$$(a, b) = \{ a + t\overrightarrow{ab} \mid 0 < t < 1 \}.$$

1.3.4 Planes

Two-dimensional affine subspaces are called *planes*. If three points a, b, c are not *collinear*, i.e., do not belong to a line, then there is a unique plane containing them, namely, the plane

$$a + \mathbb{R}\overrightarrow{ab} + \mathbb{R}\overrightarrow{ac} = \{ a + u\overrightarrow{ab} + v\overrightarrow{ac} \mid u, v, \in \mathbb{R} \}.$$

A plane contains the entire line connecting any two of its two distinct points.

1.3.5 Hyperplanes

Hyperplanes are affine subspaces of codimension 1. They are given by equations

$$a_1 x_1 + \cdots + a_n x_n = c, \tag{1.1}$$

where not all the a_i's equal 0. If we represent the hyperplane in the vector form $b + U$, where U is an $(n - 1)$-dimensional vector subspace of $\mathbb{R}^n$, then $U = \alpha^\perp$, where

$$\alpha = \begin{pmatrix} a_1 \\ \vdots \\ a_n \end{pmatrix}.$$

Two hyperplanes are either parallel or intersect along an affine subspace of dimension $n - 2$.

1.3.6 Orthogonal Projection

If A is an affine subspace in $\mathbb{AR}^n$, we define

$$A^\perp = \{ \beta \in \mathbb{R}^n \mid \alpha \cdot \beta = 0 \text{ for all } \alpha \in \overrightarrow{A} \}$$

and call it the *orthogonal complement* to A. It is easy to see that $A^\perp$ is a vector subspace and that

$$\dim A + \dim A^\perp = n.$$

Fix a point a in A; if x is a point in $\mathbb{AR}^n$, the vector $\overrightarrow{ax}$ can be uniquely decomposed as

$$\overrightarrow{ax} = \alpha + \beta, \quad \alpha \in \overrightarrow{A}, \quad \beta \in A^\perp;$$

then the point $a + \alpha$ is called the *projection* of the point x onto A, and the map

$$\mathbb{AR}^n \to A,$$

$$x \mapsto a + \alpha,$$

is called the *orthogonal projection* of $\mathbb{AR}^n$ onto A; it can be shown that it does not depend on the choice of the point $a \in A$.

1.4 Half-Spaces

If H is a hyperplane given by equation (1.1) and we denote by f the linear function

$$f(x) = a_1 x_1 + \cdots + a_n x_n - c,$$

where $x = (x_1, \ldots, x_n)$, then the hyperplane divides the affine space $\mathbb{AR}^n$ into two *open half-spaces* V^+ and V^- defined by the inequalities $f(x) > 0$ and $f(x) < 0$. The sets $\overline{V}^+$ and $\overline{V}^-$ defined by the inequalities $f(x) \geqslant 0$ and $f(x) \leqslant 0$ are called *closed half-spaces*. The half-spaces are *convex* in the following sense: if two points a and b belong to one half-space, say V^+, then the restriction of f to the segment

$$[a, b] = \{ a + t\overrightarrow{ab} \mid 0 \leqslant t \leqslant 1 \}$$

is a linear function of t that cannot take the value 0 on the segment $0 \leqslant t \leqslant 1$. Hence, if a half-space contains two points a and b, then it contains the segment $[a, b]$. Subsets in $\mathbb{AR}^n$ with this property are called *convex*.

More generally, a *curve* is an image of the segment $[0, 1]$ of the real line $\mathbb{R}$ under a continuous map from $[0, 1]$ to $\mathbb{AR}^n$. In particular, a segment $[a, b]$ is a curve, the map being

$$t \mapsto a + t\overrightarrow{ab}.$$

Two points a and b of a subset $X \subseteq \mathbb{AR}^n$ are *connected* in X if there is a curve in X containing both a and b. This is an equivalence relation, and its classes are called *connected components* of X. A set X is *path-connected* if it consists of just one connected component, that is, any two points in X can be connected by a curve belonging to X. Notice that any convex set is path-connected; in particular, half-spaces are path-connected.

If H is a hyperplane in $\mathbb{AR}^n$ then its two open half-spaces V^- and V^+ are connected components of $\mathbb{AR}^n \setminus H$. Indeed, the half-spaces V^+ and V^- are path-connected. But if we take two points $a \in V^+$ and $b \in V^-$ and consider a curve

$$\{ x(t) \mid t \in [0, 1] \} \subset \mathbb{AR}^n$$

connecting $a = x(0)$ and $b = x(1)$, then the continuous function $f(x(t))$ takes values of opposite signs at the ends of the segment $[0, 1]$ and thus must take the value 0 at some real number t_0, $0 < t_0 < 1$. But then the point $x(t_0)$ of the curve belongs to the hyperplane H.

1.5 Bases and Coordinates

Let A be an affine subspace in $\mathbb{AR}^n$ and $\dim A = k$. If $o \in A$ is an arbitrary point and $\alpha_1, \ldots, \alpha_k$ is an orthonormal basis in $\overrightarrow{A}$ then we can assign to any point $a \in A$ the coordinates $(a_1, \ldots, a_k)$ defined by the rule

$$a_i = \overrightarrow{oa} \cdot \alpha_i, \quad i = 1, \ldots, k.$$

This turns A into an affine Euclidean space of dimension k that can be identified with $\mathbb{AR}^k$. Therefore everything that we said about $\mathbb{AR}^n$ can be applied to any affine subspace of $\mathbb{AR}^n$.

We shall use a change of coordinates in the proof of the following simple fact.

Proposition 1.1. *Let a and b be two distinct points in $\mathbb{AR}^n$. The set of all points x equidistant from a and b, i.e., such that*

$$d(a, x) = d(b, x),$$

is a hyperplane normal to the segment $[a, b]$ and passing through its midpoint.

Proof. Take the midpoint o of the segment $[a, b]$ for the origin of an orthonormal coordinate system in $\mathbb{AR}^n$. Then the points a and b are represented by the vectors $\vec{oa} = \alpha$ and $\vec{ob} = -\alpha$. If x is a point with $d(a, x) = d(b, x)$, then we have, for the vector $\chi = \vec{ox}$,

$$|\chi - \alpha| = |\chi + \alpha|,$$
$$(\chi - \alpha) \cdot (\chi - \alpha) = (\chi + \alpha) \cdot (\chi + \alpha),$$
$$\chi \cdot \chi - 2\chi \cdot \alpha + \alpha \cdot \alpha = \chi \cdot \chi + 2\chi \cdot \alpha + \alpha \cdot \alpha,$$

which gives us

$$\chi \cdot \alpha = 0.$$

But this is the equation of the hyperplane normal to the vector α directed along the segment $[a, b]$. Obviously the hyperplane contains the midpoint o of the segment. $\square$

1.6 Convex Sets

Recall that a subset $X \subseteq \mathbb{AR}^n$ is *convex* if it contains for any points $x, y \in X$ the segment $[x, y]$ (Figure 1.1).

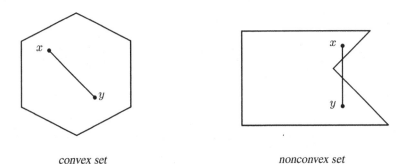

convex set   nonconvex set

Fig. 1.1. Convex and nonconvex sets.

Obviously the intersection of a collection of convex sets is convex. Every convex set is path-connected. Affine subspaces (in particular, hyperplanes) and half spaces in $\mathbb{AR}^n$ are convex. If a set X is convex then so are its closure $\overline{X}$ and interior X°. If $Y \subseteq \mathbb{AR}^n$ is a subset, its *convex hull* is defined as the intersection of all convex sets containing it; it is the smallest convex set containing Y.

Exercises

1.1. Prove that the complement of a 1-dimensional linear subspace in the 2-dimensional complex vector space $\mathbb{C}^2$ is path-connected.

1.2.* In the well known geometry textbook by Berger [Ber] the affine Euclidean spaces are defined as triples $(A, \overrightarrow{A}, \Phi)$, where $\overrightarrow{A}$ is a Euclidean vector space, A a set, and Φ a faithful simply transitive action of the additive group of $\overrightarrow{A}$ on A [Ber, vol. 1, pp. 55 and 241]. Try to understand why this is the same object as the one we discussed in this section.

1.3. Prove that affine subspaces in $\mathbb{AR}^n$ can be characterized as subsets A with the following property: for any two distinct points $a, b \in A$, the line $a + \mathbb{R}\overrightarrow{ab}$ through a and b belongs to A.

1.4. Prove that an orthogonal projection of a convex set on an affine subspace is convex.

1.5 (Henderson and Taimina [HT]). When grinding a precision flat mirror, the following method is sometimes used: Take three approximately flat pieces of glass and put pumice between the first and second pieces and grind them together. Then do the same for the second and the third pieces and then for the third and first pieces. Repeat many times and all three pieces of glass will become very accurately flat.

Why does it work? Why do we need *three* pieces of glass to achieve perfect flatness?

2

Isometries of $\mathbb{AR}^n$

In this chapter, we look at the properties of the affine Euclidean space $\mathbb{AR}^n$ as a metric space with the distance
$$d(a, b) = |\overrightarrow{ab}|.$$
An *isometry* of $\mathbb{AR}^n$ is a map s from $\mathbb{AR}^n$ onto $\mathbb{AR}^n$ that preserves distance,
$$d(sa, sb) = d(a, b) \text{ for all } a, b \in \mathbb{AR}^n.$$
We denote the group of all isometries of $\mathbb{AR}^n$ by Isom $\mathbb{AR}^n$.

We warn the reader that in this chapter, we use some elementary notions from group theory.

2.1 Fixed Points of Groups of Isometries

The following simple result will be used later in the case of finite groups of isometries.

Theorem 2.1. *Let $W <$ Isom $\mathbb{AR}^n$ be a group of isometries of $\mathbb{AR}^n$. Assume that for some point $e \in \mathbb{AR}^n$, the orbit*
$$W \cdot e = \{ we \mid w \in W \}$$
is finite. Then W fixes a point in $\mathbb{AR}^n$.

Proof. We shall use a very elementary property of triangles stated in Figure 2.1; its proof is left to the reader.

Set $E = W \cdot e$. For any point $x \in \mathbb{AR}^n$ set
$$m(x) = \max_{f \in E} d(x, f).$$

Take the point a where $m(x)$ reaches its minimum. We shall discuss the existence of the minimum a bit later (it is intuitively evident anyway). Meanwhile, we claim that the point a is unique, which would allow us to complete the proof of the theorem.

A.V. Borovik and A. Borovik, *Mirrors and Reflections: The Geometry of Finite Reflection Groups*, Universitext, DOI 10.1007/978-0-387-79066-4_2,

*In the triangle abc the segment
cd is shorter than at least one
of the sides ac and bc.*

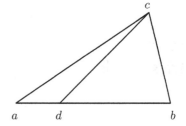

Fig. 2.1. For the proof of Theorem 2.1.

PROOF OF THE CLAIM. Indeed, if $b \neq a$ is another minimal point, take an inner point d of the segment $[a, b]$ and after that a point c such that $d(d, c) = m(d)$. We see from Figure 2.1 that for one of the points a and b, say a,

$$m(d) = d(d, c) < d(a, c) \leqslant m(a),$$

which contradicts to the minimal choice of a.

So we can return to the proof of the theorem. Since the group W permutes the points in E and preserves the distances in $\mathbb{AR}^n$, it preserves the function $m(x)$, i.e., $m(wx) = m(x)$ for all $w \in W$ and $x \in \mathbb{AR}^n$, and thus W should fix a (unique) point where the function $m(x)$ attains its minimum. The theorem is proven. However, to make the proof really watertight, we need to return to the issue of the existence of the minimum.

THE EXISTENCE OF THE MINIMUM is intuitively clear; an accurate proof consists of the following two observations. Firstly, the function $m(x)$, being the supremum of a finite number of continuous functions $d(x, f)$, is itself continuous. Secondly, we can search for the minimum not over the entire space $\mathbb{AR}^n$, but only over the set

$$\{ x \mid d(x, f) \leqslant m(a) \text{ for all } f \in E \},$$

for some $a \in \mathbb{AR}^n$. This set is closed and bounded, hence compact. But a continuous function on a compact set attains its extreme values. ◻

Notice that the proof that we have just given is a modification of a fixed-point theorem for a group acting on a space with a hyperbolic metric. J. Tits in one of his talks has attributed the proof to J.-P. Serre. An alternative (and more traditional) proof can be found in Exercise 2.2 on page 15.

2.2 Structure of Isom $\mathbb{AR}^n$

2.2.1 Translations

For every vector $\alpha \in \mathbb{R}^n$ one can define the map

$$t_\alpha : \mathbb{AR}^n \longrightarrow \mathbb{AR}^n,$$
$$a \mapsto a + \alpha.$$

The map t_α is an isometry of $\mathbb{AR}^n$; it is called *translation through the vector* α. Translations of $\mathbb{AR}^n$ form a commutative group that is obviously isomorphic to the additive group of the vector space $\mathbb{R}^n$; we shall denote it by the same symbol $\mathbb{R}^n$ as the vector space.

2.2.2 Orthogonal Transformations

When we fix an orthonormal coordinate system in $\mathbb{AR}^n$ with the origin o, a point $a \in \mathbb{AR}^n$ can be identified with its *position vector* $\alpha = \overrightarrow{oa}$. This allows us to identify $\mathbb{AR}^n$ and $\mathbb{R}^n$. Every orthogonal linear transformation w of the Euclidean vector space $\mathbb{R}^n$ can be treated as a transformation of the affine space $\mathbb{AR}^n$. Moreover, this transformation is an isometry, because by the definition of an orthogonal transformation w,

$$w\alpha \cdot w\alpha = \alpha \cdot \alpha;$$

hence $|w\alpha| = |\alpha|$ for all $\alpha \in \mathbb{R}^n$. Therefore we have, for $\alpha = \overrightarrow{oa}$ and $\beta = \overrightarrow{ob}$,

$$d(wa, wb) = |w\beta - w\alpha| = |w(\beta - \alpha)| = |\beta - \alpha| = d(a, b).$$

The group of all orthogonal linear transformations of $\mathbb{R}^n$ is called the *orthogonal group* and denoted by $\mathbb{O}_n$.

Theorem 2.2. *The group of all isometries of* $\mathbb{AR}^n$ *which fix the point o coincides with the orthogonal group* $\mathbb{O}_n$.

Proof. Let s be an isometry of $\mathbb{AR}^n$ that fixes the origin o. We have to prove that when we treat s as a map from $\mathbb{R}^n$ to $\mathbb{R}^n$, the following conditions are satisfied: for all $\alpha, \beta \in \mathbb{R}^n$,

- $s(k\alpha) = k \cdot s\alpha$ for any constant $k \in \mathbb{R}$;
- $s(\alpha + \beta) = s\alpha + s\beta$;
- $s\alpha \cdot s\beta = \alpha \cdot \beta$.

If a and b are two points in $\mathbb{AR}^n$, then by Exercise 2.3, the segment $[a, b]$ can be characterized as the set of all points x such that

$$d(a, b) = d(a, x) + d(x, b).$$

So the terminal point a' of the vector $c\alpha$ for $k > 1$ is the only point satisfying the conditions

$$d(o, a') = k \cdot d(0, a) \quad \text{and} \quad d(o, a) + d(a, a') = d(o, a').$$

If now $sa = b$, then since the isometry s preserves distances and fixes the origin o, the point $b' = sa'$ is the only point in $\mathbb{AR}^n$ satisfying

$$d(o, b') = k \cdot d(0, b) \quad \text{and} \quad d(o, b) + d(b, b') = d(o, b').$$

Hence $s \cdot k\alpha = \overrightarrow{ob'} = k\beta = k \cdot s\alpha$ for $k > 0$. The cases $k \leqslant 0$ and $0 < k \leqslant 1$ require only minor adjustments in the above proof and are left to the reader as an exercise. Thus s preserves multiplication by scalars.

The additivity of s, i.e., the property

$$s(\alpha + \beta) = s\alpha + s\beta,$$

follows, in an analogous way, from the observation that the vector $\delta = \alpha + \beta$ can be constructed in two steps: starting with the terminal points a and b of the vectors α and β, we first find the midpoint of the segment $[a, b]$ as the unique point c such that

$$d(a, c) = d(c, b) = \frac{1}{2} d(a, b),$$

and then set $\delta = 2\overrightarrow{oc}$. A detailed justification of this construction is left to the reader as an exercise.

Since s preserves distances, it preserves lengths of the vectors. But from $|s\alpha| = |\alpha|$ it follows that

$$s\alpha \cdot s\alpha = \alpha \cdot \alpha$$

for all $\alpha \in \mathbb{R}^n$. Now we apply the additivity of s and observe that

$$\begin{aligned}
(\alpha + \beta) \cdot (\alpha + \beta) &= s(\alpha + \beta) \cdot s(\alpha + \beta) \\
&= (s\alpha + s\beta) \cdot (s\alpha + s\beta) \\
&= s\alpha \cdot s\alpha + 2s\alpha \cdot s\beta + s\beta \cdot s\beta \\
&= \alpha \cdot \alpha + 2s\alpha \cdot s\beta + \beta \cdot \beta.
\end{aligned}$$

On the other hand,

$$(\alpha + \beta) \cdot (\alpha + \beta) = \alpha \cdot \alpha + 2\alpha \cdot \beta + \beta \cdot \beta.$$

Comparing these two equations, we see that

$$2s\alpha \cdot s\beta = 2\alpha \cdot \beta$$

and

$$s\alpha \cdot s\beta = \alpha \cdot \beta.$$

$\square$

Theorem 2.3. *Every isometry of a real affine Euclidean space $\mathbb{AR}^n$ is a composition of a translation and an orthogonal transformation. The group* $\mathrm{Isom}\,\mathbb{AR}^n$ *of all isometries of $\mathbb{AR}^n$ is a semidirect product of the group $\mathbb{R}^n$ of all translations and the orthogonal group $\mathbb{O}_n$,*

$$\mathrm{Isom}\,\mathbb{AR}^n = \mathbb{R}^n \rtimes \mathbb{O}_n.$$

This means that we have the following decomposition of $\mathrm{Isom}\,\mathbb{AR}^n$:

$$\mathrm{Isom}\,\mathbb{AR}^n = \mathbb{R}^n \cdot \mathbb{O}_n, \quad \mathbb{R}^n \triangleleft \mathrm{Isom}\,\mathbb{AR}^n, \quad \text{and} \quad \mathbb{R}^n \cap \mathbb{O}_n = 1.$$

Proof. The proof is an almost immediate corollary of the previous result. Indeed, if $w \in \text{Isom } \mathbb{AR}^n$ is an arbitrary isometry, take the translation $t = t_\alpha$ through the position vector $\alpha = \overrightarrow{o, wo}$ of the point wo. Then $to = wo$ and $o = t^{-1}wo$. Thus the map $s = t^{-1}w$ is an isometry of $\mathbb{AR}^n$ that fixes the origin o and, by Theorem 2.2, belongs to $\mathbb{O}_n$. Hence $w = ts$ and

$$\text{Isom } \mathbb{AR}^n = \mathbb{R}^n \mathbb{O}_n.$$

Obviously $\mathbb{R}^n \cap \mathbb{O}_n = 1$, and we need to check only that $\mathbb{R}^n \lhd \text{Isom } \mathbb{AR}^n$. But this follows from the observation that for any orthogonal transformation s,

$$st_\alpha s^{-1} = t_{s\alpha}$$

(Exercise 2.5), and consequently we have, for any isometry $w = ts$ with $t \in \mathbb{R}^n$ and $s \in \mathbb{O}_n$,

$$wt_\alpha w^{-1} = ts \cdot t_\alpha \cdot s^{-1}t^{-1} = t \cdot t_{s\alpha} \cdot t^{-1} = t_{s\alpha} \in \mathbb{R}^n.$$

Given an isometry $w = ts$ of $\mathbb{AR}^n$ with $t \in \mathbb{R}^n$ and $s \in \mathbb{O}_n$, we say that w *preserves orientation* if $\det s = +1$, and *changes orientation* if $\det s = -1$.

Exercises

2.1. Prove the property of triangles in $\mathbb{AR}^2$ stated in Figure 2.1.

2.2.* Barycenter. There is a more traditional approach to Theorem 2.1. If

$$F = \{ f_1, \ldots, f_k \}$$

is a finite set of points in $\mathbb{AR}^n$, its *barycenter* b is a point defined by the condition

$$\sum_{j=1}^{k} \overrightarrow{bf_j} = 0.$$

1. Prove that a finite set F has a unique barycenter.
2. Further, prove that the barycenter b is the point where the function

$$M(x) = \sum_{j=1}^{k} d(x, f_j)^2$$

takes its minimal value. In particular, if the set F is invariant under the action of a group W of isometries, then W fixes the barycenter b.

2.3. If a and b are two points in $\mathbb{AR}^n$, then the segment $[a, b]$ can be characterized as the set of all points x such that

$$d(a, b) = d(a, x) + d(x, b).$$

2.4. Draw a diagram illustrating the construction of $\alpha + \beta$ in the proof of Theorem 2.2, and fill in the details of the proof.

2.5. Prove that if t_α is a translation through the vector α and s is an orthogonal transformation then

$$st_\alpha s^{-1} = t_{s\alpha}.$$

2.6. Prove the following generalization of Theorem 2.1: if a group $W <$ Isom $\mathbb{AR}^n$ has a bounded orbit on $\mathbb{AR}^n$ then W fixes a point.

ELATIONS. A map $f : \mathbb{AR}^b \longrightarrow \mathbb{AR}^n$ is called an *elation* if there is a constant k such that for all $a, b \in \mathbb{AR}^n$,

$$d(f(a), f(b)) = kd(a, b).$$

An isometry is a special case $k = 1$ of an elation. The constant k is called the *coefficient* of the elation f.

2.7. An elation of $\mathbb{AR}^n$ with the coefficient k is the composition of a translation, an orthogonal transformation, and a map of the form

$$\mathbb{R}^n \longrightarrow \mathbb{R}^n,$$
$$\alpha \mapsto k\alpha.$$

2.8.* Prove that an elation of $\mathbb{AR}^n$ preserves angles: if it sends points a, b, c to the points a', b', c', respectively, then $\angle abc = \angle a'b'c'$.

2.9.* Prove that elations can be characterized as maps from $\mathbb{AR}^n$ onto $\mathbb{AR}^n$ that send straight lines to straight lines and preserveve perpendicularity.

2.10. The group of all elations of $\mathbb{AR}^n$ is isomorphic to $\mathbb{R}^n \rtimes (\mathbb{O}_n \times \mathbb{R}^{>0})$, where $\mathbb{R}^{>0}$ is the group of positive real numbers with respect to multiplication.

ISOMETRIES OF $\mathbb{AR}^3$.

2.11.* EULER'S THEOREM is a classical observation: if an isometry of $\mathbb{AR}^3$ has a fixed point and preserves orientation then it is a rotation about an axis.

3

Hyperplane Arrangements

This chapter starts to deviate from the canonical stuff of undergraduate linear algebra; we briefly discuss basic properties of an arrangement of several hyperplanes in affine space—this is already a surprisingly rich structure with some beautiful and hard mathematics.

Our exposition follows the classical treatment of the subject by Bourbaki [Bou], with slight changes in terminology. All the results mentioned in this section are intuitively self-evident, at least after drawing a few simple pictures. We omit some of the proofs, which can be found in [Bou, Chap. V, §1].

3.1 Faces of a Hyperplane Arrangement

A finite set Σ of hyperplanes in the affine space $\mathbb{AR}^n$ is called a *hyperplane arrangement*. We shall call hyperplanes in Σ *walls* of Σ.

Given an arrangement Σ, the hyperplanes in Σ cut the space $\mathbb{AR}^n$ and each other into pieces called faces; see the explicit definition below. We wish to develop a terminology for the description of relative position of faces with respect to each other.

If H is a hyperplane in $\mathbb{AR}^n$, we say that two points a and b of $\mathbb{AR}^n$ are on the *same side* of H if both of them belong to the same of the two half-spaces V^+, V^- determined by H; a and b are *similarly positioned* with respect to H if both of them belong simultaneously to either V^+, H, or V^-.

Let Σ be a finite set of hyperplanes in $\mathbb{AR}^n$. If a and b are points in $\mathbb{AR}^n$, we shall say that a and b are *similarly positioned* with respect to Σ and write $a \sim b$ if a and b are similarly positioned with respect to every hyperplane $H \in \Sigma$. Obviously $\sim$ is an equivalence relation. Its equivalence classes are called *faces* of the hyperplane arrangement Σ (Figure 3.1). Since Σ is finite, it has only finitely many faces. We emphasize that faces are *disjoint*; distinct faces have no points in common.

It easily follows from the definition that if F is a face and a hyperplane $H \in \Sigma$ contains a point in F then H contains F. The intersection L of all hyperplanes in Σ that contain F is an affine subspace. It is called the *support* of F. The *dimension* of F is the dimension of its support L.

A.V. Borovik and A. Borovik, *Mirrors and Reflections: The Geometry of Finite Reflection Groups*, Universitext, DOI 10.1007/978-0-387-79066-4_3, © Springer Science+Business Media, LLC 2010

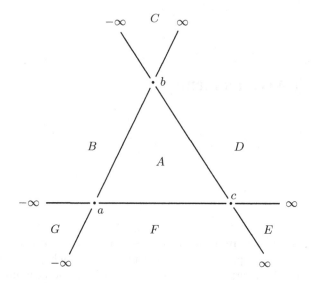

Fig. 3.1. Three lines in general position (i.e., no two lines are parallel and three lines do not intersect in one point) divide the plane into seven open faces $A, \ldots, G$ (chambers), nine 1-dimensional faces (edges) $(-\infty, a), (a, b), \ldots, (c, \infty)$, and three 0-dimensional faces (vertices) a, b, c. Notice that 1-dimensional faces are open intervals.

Topological properties of faces are described by the following result.

Proposition 3.1. *In this notation,*

- *F is an open convex subset of the affine space L.*
- *The boundary of F is the union of some set of faces of strictly smaller dimension.*
- *If F and F' are faces with equal closures, $\overline{F} = \overline{F'}$, then $F = F'$.*

3.2 Chambers

By definition, *chambers* are faces of Σ that are not contained in any hyperplane of Σ. Also chambers can be defined, in an equivalent way, as connected components of

$$\mathbb{AR}^n \smallsetminus \bigcup_{H \in \Sigma} H.$$

Chambers are open convex subsets of $\mathbb{AR}^n$. A *panel* or *facet* of a chamber C is a face of dimension $n - 1$ on the boundary of C. It follows from the definition that a panel P belongs to a unique hyperplane $H \in \Sigma$, called a *wall* of the chamber C.

Proposition 3.2. *Let C and C' be two chambers. The following conditions are equivalent:*

- C and C' are separated by just one hyperplane in Σ.
- C and C' have a panel in common.
- C and C' have a unique panel in common.

Lemma 3.3. *Let C and C' be distinct chambers and P their common panel. Then*

(a) *the wall H that contains P is the only wall with a nontrivial intersection with the set $C \cup P \cup C'$, and*

(b) *$C \cup P \cup C'$ is a convex open set.*

Proof. The set $C \cup P \cup C'$ is a connected component of what is left after deleting from V all hyperplanes from Σ but H. Therefore H is the only wall in Σ that intersects $C \cup P \cup C'$. Moreover, $C \cup P \cup C'$ is the intersection of open half-spaces and hence is convex. $\square$

3.3 Galleries

We say that chambers C and C' are *adjacent* if they have a panel in common. Notice that a chamber is adjacent to itself. A *gallery* Γ is a sequence $C_0, C_1, \ldots, C_l$ of chambers such that C_i and C_{i-1} are adjacent, for all $i = 1, \ldots, l$. The number l is called the *length* of the gallery. We say that C_0 and C_l are *connected* by the gallery Γ and that C_0 and C_l are the *endpoints* of Γ. A gallery is *geodesic* if it has the minimal length among all galleries connecting its endpoints. The *distance* $\mathrm{gd}(C, D)$ between the chambers C and D is the length of a geodesic gallery connecting them.

Proposition 3.4. *Any two chambers of Σ can be connected by a gallery, and hence by a geodesic gallery. The distance $\mathrm{gd}(D, C)$ between the chambers C and D equals the number of hyperplanes in Σ that separate C from D.*

Proof. Assume that C and D are separated by m hyperplanes in Σ. Select two points $c \in C$ and $d \in D$ such that the segment $[c, d]$ does not intersect any $(n - 2)$-dimensional face of Σ. Then the chambers that are intersected by the segment $[c, d]$ form a gallery connecting C and D, and it is easy to see that its length is m. To prove that $m = \mathrm{gd}(C, D)$, consider an arbitrary gallery $C_0, \ldots, C_l$ connecting $C = C_0$ and $D = C_l$. We may assume without loss of generality that successive chambers C_{i-1} and C_i are distinct for all $i = 1, \ldots, l$. For each $i = 0, 1, \ldots, l$, choose a point $c_i \in C_i$. The union

$$[c_0, c_1] \cup [c_1, c_2] \cup \cdots \cup [c_{l-1}, c_l]$$

is connected, and by the connectedness argument each wall H that separates C and D has to intersect one of the segments $[c_{i-1}, c_i]$. Let P be the common panel of C_{i-1} and C_i. By virtue of Lemma 3.3(a), $[c_{i-1}, c_i] \subset C_{i-1} \cup P \cup C_i$ and H has a nontrivial intersection with $C_{i-1} \cup P \cup C_i$. But then, in view of Lemma 3.3(b), H contains the panel P. Therefore each of m walls separating C from D contains the common panel of a different pair (C_{i-1}, C_i) of adjacent chambers. It is obvious now that $l \geqslant m$. $\square$

As a byproduct of this proof, we have another useful result.

Lemma 3.5. *Assume that the endpoints of the gallery $C_0, C_1, \ldots, C_l$ lie on opposite sides of the wall H. Then, for some $i = 1, \ldots, l$, the wall H contains the common panel of consecutive chambers C_{i-1} and C_i.*

We shall say in this situation that the wall H *intersects* the gallery $C_0, \ldots, C_l$.

Another corollary of Proposition 3.4 is the following characterization of geodesic galleries.

Proposition 3.6. *A gallery is geodesic if and only if it intersects each wall at most once.*

Corollary 3.7. *Let*
$$C = C_0, C_1, \ldots, C_l = D$$
be a geodesic gallery. Let C_i and C_{i+1} be consecutive chambers and H the wall separating them. Then the endpoints C and D of the gallery lie on opposite sides of H.

The following elementary property of distance $\mathrm{gd}(\,,\,)$ will be very useful in the sequel.

Proposition 3.8. *Let D and E be two distinct adjacent chambers and H the wall separating them. Let C be a chamber, and assume that the chambers C and D lie on the same side of H. Then*

$$\mathrm{gd}(C, E) = \mathrm{gd}(C, D) + 1.$$

Proof. The proof is left to the reader as an exercise (Exercise 3.5). □

3.4 Polyhedra

A *polyhedral set*, or *polyhedron*, in $\mathbb{AR}^n$ is the intersection of a finite number of closed half-spaces. Since half-spaces are convex, every polyhedron is convex. Bounded polyhedra are called *polytopes* (Figure 3.2).

Let Δ be a polyhedron represented as the intersection of closed half-spaces $X_1, \ldots, X_m$ bounded by the hyperplanes $H_1, \ldots, H_m$. Consider the hyperplane configuration $\Sigma = \{ H_1, \ldots, H_m \}$. If F is a face of Σ and has a point in common with Δ then F is contained in Δ. Thus Δ is a union of faces (Figure 3.3). Actually it can be shown that Δ is the closure of exactly one face of Σ.

0-dimensional faces of Δ are called *vertices*, and the 1-dimensional faces are called *edges*.

The following result is probably the most important theorem about polytopes.

Theorem 3.9. *A polytope is the convex hull of its vertices. Conversely, given a finite set E of points in $\mathbb{AR}^n$, their convex hull is a polytope whose vertices belong to E.*

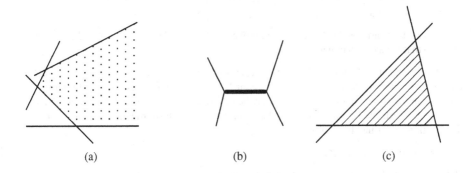

(a)   (b)   (c)

Fig. 3.2. Polyhedra can be unbounded (a) or without interior points (b). In some books the term "polytope" is reserved for bounded polyhedra with interior points (c); we prefer to use it for all bounded polyhedra, so that (b) is a polytope in our sense.

Fig. 3.3. A polyhedron is the union of its faces.

As R. T. Rockafellar characterized it [Roc, p. 171],

> This classical result is an outstanding example of a fact which is completely obvious to geometric intuition, but which wields important algebraic content and is not trivial to prove.

We hope this quotation is a sufficient justification for our decision not to include the proof of the theorem in our book.

Exercises

LINES ON THE PLANE. Several lines on the plane already make a remarkably rich structure with many beautiful properties. Any serious course in geometry should start with the detailed study of this basic configuration. A few problems illustrate this point; they are included here with the aim to help the novice reader to develop geometric intuition.

3.1. Prove that in the plane $\mathbb{AR}^2$, n lines in general position (i.e., no lines are parallel and no three intersect in one point) divide the plane into

$$1 + (1 + 2 + \cdots + n) = \frac{1}{2}(n^2 + n + 2)$$

chambers. How many of these chambers are unbounded? Also, find the numbers of 1- and 0-dimensional faces.

3.2. Given a line arrangement in the plane, prove that the chambers can be colored black and white so that adjacent chambers have different colors.

3.3. Four straight lines in the plane are in general position (that is, no two lines are parallel and no three lines meet at the same point). Four hikers walk along these lines with constant speeds (but the speeds of different hikers may be different). It is known that the first hiker met the second, third, and fourth ones, while the second hiker met the third and fourth ones. Prove that the third hiker met the fourth one.

GALLERIES AND DISTANCE.

3.4. Prove that distance gd(,) on the set of chambers of a hyperplane arrangement satisfies the triangle inequality:

$$gd(C, D) + gd(D, E) \geqslant gd(C, E).$$

3.5. Prove Proposition 3.8.

TETRAHEDRA AND n-SIMPLICES.

3.6. Let Δ be a tetrahedron in $\mathbb{AR}^3$ and Σ the arrangement formed by the planes containing facets of Δ. Make a sketch analogous to Figure 3.1. Find the number of chambers of Σ. Can you see a natural correspondence between chambers of Σ and faces of Δ?

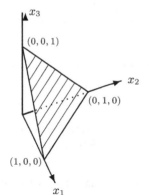

The regular 2-*simplex is the set of solutions of the system of simultaneous inequalities and equation*

$$x_1 + x_2 + x_3 = 0,$$

$$x_1 \geqslant 0, \ x_2 \geqslant 0, \ x_3 \geqslant 0.$$

We see that it is an equilateral triangle.

Fig. 3.4. The regular 2-simplex.

3.7. The previous exercise can be generalized to the case of n dimensions in the following way. By definition, the *regular n-simplex* is the set of solutions of the system of simultaneous inequalities and equation

$$x_1 + \cdots + x_n + x_{n+1} = 1,$$
$$x_1 \geqslant 0,$$
$$\vdots$$
$$x_{n+1} \geqslant 0$$

(Figure 3.4). It is the polytope in the n-dimensional affine subspace A with the equation

$$x_1 + \cdots + x_{n+1} = 1$$

bounded by the coordinate hyperplanes $x_i = 0, i = 1, \ldots, n+1$ (Figure 3.4). Prove that these hyperplanes cut A into $2^{n+1} - 1$ chambers.

GROUPS OF SYMMETRIES.

3.8. If $\Delta \subset \mathbb{AR}^n$, the *group of symmetries* Sym Δ of the set Δ consists of all isometries of $\mathbb{AR}^n$ that map Δ onto Δ. Give examples of polytopes Δ in $\mathbb{AR}^3$ such that

1. Sym Δ acts transitively on the set of vertices of Δ but is intransitive on the set of faces;
2. Sym Δ acts transitively on the set of faces of Δ but is intransitive on the set of vertices;
3. Sym Δ is transitive on the set of edges of Δ but is intransitive on the set of faces.

3.9. Prove that the symmetry group of a polytope is finite.

3.10. Construct a polyhedron with an infinite symmetry group.

4

Polyhedral Cones

In this book, we shall deal mostly with arrangements of hyperplanes in $\mathbb{AR}^n$ that pass through a common point o. In that case, the hyperplanes cut the space into *polyhedral cones*. The theory of polyhedral cones is closely related to the solution of systems of linear inequalities in several variables and is traditionally treated as part of linear programming. We list here only the most basic properties of polyhedral cones.

4.1 Finitely Generated Cones

4.1.1 Cones

A *cone* in $\mathbb{R}^n$ is a subset C closed under addition and positive scalar multiplication, that is, $\alpha + \beta \in \Gamma$ and $k\alpha \in C$ for any $\alpha, \beta \in C$ and scalar $k > 0$. Linear subspaces and half-spaces of $\mathbb{R}^n$ are cones. Every cone is convex, since it contains, for any two points of the cone α and β, the segment

$$[\alpha, \beta] = \{ (1 - t)\alpha + t\beta \mid 0 \leqslant t \leqslant 1 \}.$$

A cone does not necessarily contain the zero vector 0; this is the case, for example, for the positive quadrant C in $\mathbb{R}^2$,

$$C = \left\{ \begin{pmatrix} x \\ y \end{pmatrix} \in \mathbb{R}^2 \ \middle| \ x > 0, \, y > 0 \right\}.$$

However, we can always add to a cone the origin O of $\mathbb{R}^n$: if C is a cone then so is $C \cup \{O\}$. It can be shown that if C is a cone then so are its topological closure $\overline{C}$ and interior C°. The intersection of a collection of cones is either a cone or the empty set.

The cone C *spanned* or *generated* by a set of vectors Π is the set of all nonnegative linear combinations of vectors from Π,

$$C = \{ a_1\alpha_1 + \cdots + a_m\alpha_m \mid m \in \mathbb{N}, \, \alpha_i \in \Pi, \, a_i \geqslant 0 \}.$$

A.V. Borovik and A. Borovik, *Mirrors and Reflections: The Geometry of Finite Reflection Groups*, Universitext, DOI 10.1007/978-0-387-79066-4_4,
© Springer Science+Business Media, LLC 2010

Notice that the zero vector 0 belongs to C. If the cone C is spanned by a finite set Π then C is called *finitely generated* and the set Π is a *system of generators* for C. A cone is *polyhedral* if it is a polyhedral set, i.e., the intersection of a finite number of closed half spaces, with the origin O belonging to the bounding hyperplane of each of these subspaces.

The following important result can be found in most books on linear programming. In this book we shall prove only a very restricted special case, Proposition 4.6 below.

Theorem 4.1. *A cone is finitely generated if and only if it is polyhedral.*

4.1.2 Extreme Vectors and Edges

We shall call a set of vectors Π *positive* if for some linear function

$$f : \mathbb{R}^n \longrightarrow \mathbb{R},$$

$f(\rho) > 0$ for all $\rho \in \Pi \smallsetminus \{0\}$. This is equivalent to saying that the set $\Pi \smallsetminus \{0\}$ of nonzero vectors in Π is contained in an open half-space. The following property of positive sets of vectors is fairly obvious.

Lemma 4.2. *If $\alpha_1, \ldots, \alpha_m$ are nonzero vectors in a positive set Π and*

$$a_1\alpha_1 + \cdots + a_m\alpha_m = 0, \quad \text{where all} \quad a_i \geqslant 0,$$

then $a_i = 0$ for all $i = 1, \ldots, m$.

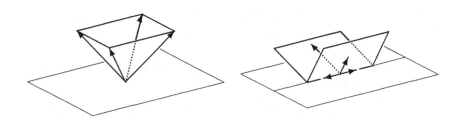

a pointed finitely generated cone   a nonpointed finitely generated cone

Fig. 4.1. Pointed and nonpointed cones.

Positive cones are usually called *pointed* cones (Figure 4.1).

Let C be a cone in $\mathbb{R}^n$. We shall say that a vector $\epsilon \in C$ is *extreme* or *simple* in C if it cannot be represented as a positive linear combination which involves vectors

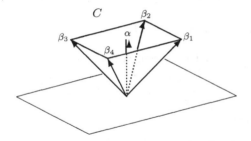

Fig. 4.2. Extreme and nonextreme vectors

α *is a nonextreme vector in the cone* C *generated by extreme* (*or simple*) *vectors* $\beta_1, \beta_2, \beta_3, \beta_4$ *directed along the edges of* C.

in C noncolinear to ϵ, i.e., if it follows from $\epsilon = c_1\gamma_1 + \cdots + c_m\gamma_m$, where $\gamma_i \in C$ and $c_i > 0$, that $m = 1$ and $\epsilon = c_1\gamma_1$. Notice that it immediately follows from the definition that if ϵ is an extreme vector and Π a system of generators in C then Π contains a vector $k\epsilon$ collinear to ϵ.

Extreme vectors in a polyhedral cone $C \subset \mathbb{R}^2$ or $\mathbb{R}^3$ have the most natural geometric interpretation: these are vectors directed along the edges of C. We prefer to take this property for the definition of an edge: if ϵ is an extreme vector in a polyhedral cone C then the cone $C \cap \mathbb{R}\epsilon$ is called an *edge* of C; see Figure 4.2.

4.2 Simple Systems of Generators

A finite system Π of generators in a cone C is said to be *simple* if it consists of simple vectors and no two distinct vectors in Π are collinear. It follows from the definition of an extreme vector that any two simple systems Π and Π' in C contain an equal number of vectors; moreover, every vector in Π is collinear to some vector in Π', and vice versa.

Proposition 4.3. *Let Π be a finite positive set of vectors and C the cone it generates. Assume also that Π contains no collinear vectors, that is, $\alpha = k\beta$ for distinct vectors $\alpha, \beta \in \Pi$ and $k \in \mathbb{R}$ implies $k = 0$. Then Π contains a (unique) simple system of generators.*

In geometric terms this means that a finitely generated pointed cone has finitely many edges and is generated by a system of vectors directed along the edges, one vector from each edge.

Proof. We shall prove the following claim, which makes the statement of the lemma obvious.

A nonextreme vector can be removed from any generating set for a pointed cone C. In more precise terms, if the vectors $\alpha, \beta_1, \ldots, \beta_k$ of Π generate C and α is not an extreme vector then the vectors $\beta_1, \ldots, \beta_k$ still generate C.

Proof of the claim. Let

$$\Pi = \{\,\alpha, \beta_1, \ldots, \beta_k, \gamma_1, \ldots, \gamma_l\,\},$$

where no γ_j is collinear with α. Since α is not an extreme vector,

$$\alpha = \sum_{i=1}^{k} b_i \beta_i + \sum_{j=1}^{l} c_j \gamma_j, \quad b_i \geqslant 0, \quad c_l \geqslant 0.$$

Also, since the vectors $\alpha, \beta_1, \ldots, \beta_k$ generate the cone C,

$$\gamma_j = d_j \alpha + \sum_{i=1}^{k} f_{ji} \beta_i, \quad d_j \geqslant 0, \quad f_{ji} \geqslant 0.$$

Substituting γ_i from the latter equations into the former, we have, after a simple rearrangement,

$$\left(1 - \sum_{j=1}^{l} c_j d_j \right) \alpha = \sum_{i=1}^{k} \left(b_i + \sum_{j=1}^{l} c_j f_{ji} \right) \beta_i.$$

The vector α and the vector on the right-hand side of this equation both lie in the same open half-space; therefore, in view of Lemma 4.2,

$$1 - \sum_{j=1}^{l} c_j d_j > 0,$$

and

$$\alpha = \frac{1}{1 - \sum_j c_j d_j} \sum_{i=1}^{k} \left(b_i + \sum_{j=1}^{l} c_j f_{ji} \right) \beta_i$$

expresses α as a nonnegative linear combination of the β_i's. Since the vectors

$$\alpha, \beta_1, \ldots, \beta_k$$

generate C, the vectors $\beta_1, \ldots, \beta_k$ also generate C.   □

The following simple lemma has an even simpler geometric interpretation: the plane passing through two edges of a cone cuts it in the cone spanned by these two edges; see Figure 4.3.

Lemma 4.4. *Let α and β be two distinct extreme vectors in a finitely generated cone C. Let P be the plane (2-dimensional vector subspace) spanned by α and β. Then $C_0 = C \cap P$ is the cone in P spanned by α and β.*

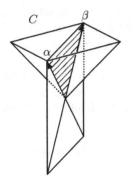

The intersection of a cone C with the plane spanned by two simple vectors α and β is the cone generated by α and β.

Fig. 4.3. For the proof of Lemma 4.4.

Proof. Assume the contrary; let $\gamma \in C_0$ be a vector that does not belong to the cone spanned by α and β. Since α and β form a basis in the vector space P,

$$\gamma = a'\alpha + b'\beta,$$

and by our assumption one of the coefficients a' and b' is negative. We can assume without loss of generality that $b' < 0$.

Let $\alpha, \beta, \gamma_1, \ldots, \gamma_m$ be the simple system in C. Since $\gamma \in C$,

$$\gamma = a\alpha + b\beta + c_1\gamma_1 + \cdots + c_m\gamma_m,$$

where all the coefficients $a, b, c_1, \ldots, c_m$ are nonnegative. Comparing the two expressions for γ, we have

$$(a - a')\alpha + (b - b')\beta + c_1\gamma_1 + \cdots + c_m\gamma_m = 0.$$

Notice that $b - b' > 0$; if $a - a' \geqslant 0$ then we get a contradiction to the assumption that the cone C is pointed. Therefore $a - a' < 0$, and

$$\alpha = \frac{1}{a' - a}\left((b - b')\beta + c_1\gamma_1 + \cdots + c_m\gamma_m\right)$$

expresses α as a nonnegative linear combination of the rest of the simple system. This contradiction proves the lemma. $\qquad\square$

4.3 Duality

If C is a cone, the *dual cone* C^* is the set

$$C^* = \{\chi \in \mathbb{R}^n \mid \chi \cdot \gamma \leqslant 0 \text{ for all } \gamma \in C\}.$$

It immediately follows from this definition that the set C^* is closed with respect to addition and multiplication by positive scalars, so the name "cone" for it is justified. Also, the dual cone C^*, being the intersection of closed half-spaces $\chi \cdot \gamma \leqslant 0$, is closed in the topological sense.

The following theorem plays an extremely important role in several branches of mathematics: linear programming, functional analysis, convex geometry. We shall not use or prove it in its full generality, proving instead a simpler partial case, Proposition 4.6.

Theorem 4.5. (The duality theorem for polyhedral cones) *If C is a polyhedral cone, then so is C^*. Moreover,*

$$(C^*)^* = C.$$

Recall that polyhedral cones are closed by definition.

4.4 Duality for Simplicial Cones

A finitely generated cone $C \subset \mathbb{R}^n$ is called *simplicial* if it is spanned by n linearly independent vectors $\rho_1, \ldots, \rho_n$. Define

$$\Pi = \{\, \rho_1, \ldots, \rho_n \,\}.$$

We shall prove the duality theorem, Theorem 4.5, in the special case of simplicial cones, and obtain, in the course of the proof, very detailed information about their structure.

First of all, notice that if the cone C is generated by a finite set $\Pi = \{\, \rho_1, \ldots, \rho_n \,\}$ then the inequalities

$$\chi \cdot \gamma \leqslant 0 \quad \text{for all } \gamma \in C$$

are equivalent to

$$\chi \cdot \rho_i \leqslant 0, \quad i = 1, \ldots, n.$$

Hence the dual cone C^* is the intersection of the closed subspaces given by the inequalities

$$\chi \cdot \rho_i \leqslant 0, \quad i = 1, \ldots, n.$$

We know from linear algebra that the conditions

$$\rho_i^* \cdot \rho_j = \begin{cases} -1 & \text{if } i = j, \\ 0 & \text{if } i \neq j, \end{cases}$$

uniquely determine n linearly independent vectors $\rho_1^*, \ldots, \rho_n^*$ (see Exercises 4.3 and 4.4). We shall say that the basis

$$\Pi^* = \{\, \rho_1^*, \ldots, \rho_n^* \,\}$$

is *dual*[1] to the basis $\rho_1, \ldots, \rho_n$. If we write a vector $\chi \in \mathbb{R}^n$ in the basis Π^*,

$$\chi = y_1^* \rho_1^* + \cdots + y_n^* \rho_n^*,$$

then $\chi \cdot \rho_i = -y_i^*$ and $\chi \in C^*$ if and only if $y_i \geq 0$ for all i, which means that $\chi \in C^*$. So we have proved the following partial case of the duality theorem, illustrated by Figure 4.4.

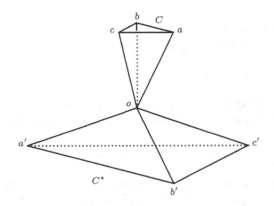

The simplicial cones C and C^* are dual to each other:

$$oa \perp b'oc', \quad ob \perp c'oa',$$
$$oc \perp a'ob', \quad oa' \perp boc,$$
$$ob' \perp coa, \quad oc' \perp aob.$$

Fig. 4.4. Dual simplicial cones.

Proposition 4.6. *If C is the simplicial cone spanned by a basis Π of $\mathbb{R}^n$ then the dual cone C^* is also simplicial and spanned by the dual basis Π^*. Applying this property to C^*, we see that $C = (C^*)^*$ is the dual cone to C^* and coincides with the intersection of the closed half-spaces*

$$\chi \cdot \rho_i^* \leq 0, \quad i = 1, \ldots, n.$$

4.5 Faces of a Simplicial Cone

Denote by H_i the hyperplane $\chi \cdot \rho_i^* = 0$. Notice that the cone C lies in one closed half-space determined by H_i. The intersection $C_k = C \cap H_k$ consists of all vectors of the form

$$\chi = y_1 \rho_1 + \cdots + y_n \rho_n$$

---

[1] We move a little bit away from the traditional terminology, since the dual basis is usually defined by the conditions

$$\rho_i^* \cdot \rho_j = \begin{cases} 1 & \text{if } i = j, \\ 0 & \text{if } i \neq j. \end{cases}$$

with nonnegative coordinates y_i, $i = 1, \ldots, n$, and zero kth coordinate, $y_k = 0$. Therefore C_k is the simplicial cone in the $(n-1)$-dimensional vector space $\chi \cdot \rho_k^* = 0$ spanned by the vectors ρ_i, $i \neq k$. The cones C_k are called *facets* or $(n-1)$-dimensional *faces* of C.

More generally, if we define $I = \{1, \ldots, n\}$ and take a subset $J \subset I$ of cardinality m, then the $(n-m)$-*dimensional face* C_J of C can be defined in two equivalent ways:

- C_J is the cone spanned by the vectors ρ_i, $i \in I \smallsetminus J$.
- $C_J = C \cap \bigcap_{j \in J} H_j$.

It follows from their definition that edges are 1-dimensional faces.

If we define the faces C_J^* in an analogous way then we have the formula

$$C_J^* = \{\chi \in C^* \mid \chi \cdot \gamma = 0 \text{ for all } \gamma \in C_{I \smallsetminus J}\}.$$

Abusing terminology, we shall say that the face C_J^* of C^* is *dual* to the face $C_{I \smallsetminus J}$ of C. This defines a one-to-one correspondence between the faces of the simplicial cone C and its dual C^*.

In particular, the edges of C are dual to facets of C^*, and the facets of C are dual to edges of C^*.

We shall use also the duality theorem for cones C spanned by $m < n$ linearly independent vectors in $\mathbb{R}^n$. The description of C^* in this case is an easy generalization of Proposition 4.6; see Exercise 4.5.

Exercises

4.1. A finite set Π of nonzero vectors in $\mathbb{R}^n$ generates a pointed cone if and only if the zero vector cannot be represented as a positive linear combination of vectors from Π, that is, the relation

$$a_1 \alpha_1 + \cdots + a_m \alpha_m = 0, \quad a_i \geqslant 0 \text{ for all } i = 1, \ldots, m,$$

implies $a_i = 0$ for all $i = 1, \ldots, m$.

4.2. Let X be an arbitrary positive set of vectors in $\mathbb{R}^n$. Prove that the set

$$X^* = \{\alpha \in \mathbb{R}^n \mid \alpha \cdot \gamma \leqslant 0\}$$

is a cone. Show next that X^* contains a nonzero vector and that X is contained in the cone $(X^*)^*$.

4.3. DUAL BASIS. Let $\epsilon_1, \ldots, \epsilon_n$ be an orthonormal basis and $\rho_1, \ldots, \rho_n$ a basis in $\mathbb{R}^n$. Form the matrix $R = (r_{ij})$ by the rule $r_{ij} = \rho_i \cdot \epsilon_j$, so that

$$\rho_i = \sum_{j=1}^n r_{ij} \epsilon_j.$$

Notice that R is a nondegenerate matrix. Let $\rho = y_1 \epsilon_1 + \cdots + y_n \epsilon_n$. For each value of i, express the system of simultaneous equations

$$(\rho, \rho_j) = \begin{cases} -1 & \text{if } i = j, \\ 0 & \text{if } i \neq j, \end{cases}$$

in matrix form and prove that it has a unique solution. This will prove the existence of the basis dual to $\rho_1, \ldots, \rho_n$.

4.4. A FORMULA FOR THE DUAL BASIS. In the notation of Exercise 4.3, prove that the dual basis $\{\rho_i^*\}$ can be determined from the formula

$$\rho_j^* = -\frac{1}{\det R} \begin{vmatrix} r_{11} & \cdots & r_{1,j-1} & \epsilon_1 & r_{1,j+1} & \cdots & r_{1,n} \\ r_{21} & \cdots & r_{2,j-1} & \epsilon_2 & r_{2,j+1} & \cdots & r_{2,n} \\ \vdots & & \vdots & \vdots & \vdots & & \vdots \\ r_{i,1} & \cdots & r_{i,j-1} & \epsilon_i & r_{i,j+1} & \cdots & r_{i,n} \\ \vdots & & \vdots & \vdots & \vdots & & \vdots \\ r_{n,1} & \cdots & r_{n,j-1} & \epsilon_n & r_{n,j+1} & \cdots & r_{n,n} \end{vmatrix}.$$

Notice that in the case $n = 3$ we arrive at the formula

$$\rho_1^* = -\frac{1}{(\rho_1, \rho_2, \rho_3)} \rho_2 \times \rho_3, \quad \rho_2^* = -\frac{1}{(\rho_1, \rho_2, \rho_3)} \rho_3 \times \rho_1, \quad \rho_3^* = -\frac{1}{(\rho_1, \rho_2, \rho_3)} \rho_1 \times \rho_2,$$

where $(\ ,\ ,\ )$ denotes the scalar triple product

$$(\rho_1, \rho_2, \rho_3) = \rho_1 \cdot (\rho_2 \times \rho_3),$$

and $\times$ the cross (or vector) product of vectors.

4.5. Let C be a cone in $\mathbb{R}^n$ spanned by a set Π of m linearly independent vectors $\rho_1, \ldots, \rho_m$, with $m < n$. Let U be the vector subspace spanned by Π.

Then C is a simplicial cone in U; we denote its dual in U by C', and set C^* to be the dual cone for C in V. Let also $\Pi' = \{\rho'_1, \ldots, \rho'_m\}$ be the basis in U dual to the basis Π. We shall use in the sequel the following properties of the cone C^*:

1. For any set $A \in \mathbb{R}^n$, define

$$A^\perp = \{\chi \in \mathbb{R}^n \mid \chi \cdot \alpha = 0 \text{ for all } \gamma \in A\}.$$

   Check that $A^\perp$ is a linear subspace of $\mathbb{R}^n$. Prove that $\dim C^\perp = n - m$.
   Hint: $C^\perp = U^\perp$.

2. C^* is the intersection of the closed half-spaces defined by the inequalities $\chi \cdot \rho_i \leqslant 0$, $i = 1, \ldots, m$.

3. $C^* = C' + C^\perp$; this set is, by definition,

$$C' + C^\perp = \{\kappa + \chi \mid \kappa \in C', \chi \in C^\perp\}.$$

4. $(C^*)^* = C$. (This extends Proposition 4.6.)

5. Let H_i and H_i^* be hyperplanes in V given by the equations $\chi \cdot \rho'_i = 0$ and $\chi \cdot \rho_i = 0$, respectively. Define $I = \{1, \ldots, m\}$ and set, for $J \subseteq I$,

$$C_J = C \cap \bigcap_{j \in J} H_j, \quad C_J^* = C^* \cap \bigcap_{j \in J} H_j^*, \quad \text{and} \quad C'_J = C' \cap \bigcap_{j \in J} H_j^*.$$

   Prove that $C_J^* = C'_J + C^\perp$.

6. The cones C_J and C_J^* are called *faces* of the cones C and C^*, respectively. Prove that there is a one-to-one correspondence between the set of k-dimensional faces of C, $k = 1, \ldots, m - 1$, and $(n - k)$-dimensional faces of C^*, defined by the rule

$$C_J^* = C \cap C_{I \smallsetminus J}^{\perp}.$$

If we treat C as its own m-dimensional face $C_\emptyset$, then it corresponds to $C_I^* = C^\perp$.

Part II

Mirrors, Reflections, Roots

5

Mirrors and Reflections

This chapter bears the same title as the book; this is where its main characters first appear on the scene.

We define *reflection* in an affine real Euclidean space $A\mathbb{R}^n$ to be a nonidentity isometry s that fixes all points of some affine hyperplane H of $A\mathbb{R}^n$. The hyperplane H is called the *mirror* of the reflection s and denoted by H_s. Conversely, the reflection s will be sometimes denoted by $s = s_H$. This notation and terminology are justified, since we shall soon see that a reflection is uniquely determined by its mirror.

Lemma 5.1. *If s is a reflection with the mirror H, then for any point $\alpha \in A\mathbb{R}^n$,*

- *the segment $[s\alpha, \alpha]$ is normal to H and H intersects the segment in its midpoint;*
- *H is the set of points fixed by s;*
- *s is an involutary transformation, that is, $s^2 = 1$.*

In particular, the reflection s is uniquely determined by its mirror H, and vice versa.

Note that a nonidentity element g of a group G is called an *involution* if it has order 2. Hence s is an involution. In particular, $s^{-1} = s$.

Proof. Choose some point of H for the origin O of an orthonormal coordinate system, and identify the affine space $A\mathbb{R}^n$ with the underlying real Euclidean vector space $\mathbb{R}^n$. Then, by Theorem 2.2, s can be identified with an orthogonal transformation of $\mathbb{R}^n$. Since s fixes all points in H, it has at least $n - 1$ eigenvalues 1, and since s is not the identity, the only possibility for the remaining eigenvalue is -1. In particular, s is diagonalizable and has order 2, that is, $s^2 = 1$ and $s \neq 1$. It also follows from here that H is the set of all points fixed by s.

If now we consider the vector $s\alpha - \alpha$ directed along the segment $[s\alpha, \alpha]$, then

$$s(s\alpha - \alpha) = s^2\alpha - s\alpha = \alpha - s\alpha,$$

which means that the vector $s\alpha - \alpha$ is an eigenvector of s for the eigenvalue -1. Hence the segment $[s\alpha, \alpha]$ is normal to H. Its midpoint $\frac{1}{2}(s\alpha + \alpha)$ is s-invariant, since

A.V. Borovik and A. Borovik, *Mirrors and Reflections: The Geometry of Finite Reflection Groups*, Universitext, DOI 10.1007/978-0-387-79066-4_5, © Springer Science+Business Media, LLC 2010

$$s\frac{1}{2}(s\alpha + \alpha) = \frac{1}{2}(s^2\alpha + s\alpha) = \frac{1}{2}(s\alpha + \alpha),$$

hence belongs to H.   □

In the course of the proof of the previous lemma we have also shown the following

Lemma 5.2. *Reflections in $\mathbb{AR}^n$ that fix the origin o are exactly the orthogonal transformations of $\mathbb{R}^n$ with $n - 1$ eigenvalues 1 and one eigenvalue -1; their mirrors are eigenspaces for the eigenvalue 1.*

We say that the points $s\alpha$ and α are *symmetric* in H. If $X \subset A$ then the set sX is called the *reflection* or the *mirror image* of the set X in the mirror H.

Lemma 5.3. *If t is an isometry of $\mathbb{AR}^n$, s the reflection in the mirror H, and s' the reflection in tH, then*

$$s' = tst^{-1}.$$

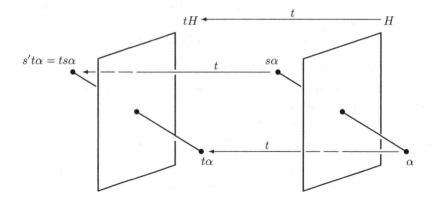

Fig. 5.1. For the proof of Lemma 5.3: If s is the reflection in the mirror H and t is an isometry then the reflection s' in the mirror tH can be found from the condition $s't = ts$, hence $s' = tst^{-1}$

Proof. See Figure 5.1. Alternatively, we may argue as follows.

We need only show that tst^{-1} is a nonidentity isometry that fixes tH. Since tst^{-1} is a composition of isometries, it is clearly an isometry. If $\alpha \in tH$, then $t^{-1}\alpha \in H$; hence s fixes $t^{-1}\alpha$, and hence $tst^{-1}\alpha = \alpha$. If $\alpha \notin tH$, then $t^{-1}\alpha \notin H$; hence s does not fix $t^{-1}\alpha$, and hence $tst^{-1}\alpha \neq \alpha$.   □

Exercises

5.1. Prove that every 2×2 orthogonal matrix A over $\mathbb{R}$ can be written in one of the forms

$$\begin{pmatrix} \cos\theta & -\sin\theta \\ \sin\theta & \cos\theta \end{pmatrix}, \quad \begin{pmatrix} \cos\theta & \sin\theta \\ \sin\theta & -\cos\theta \end{pmatrix},$$

depending on whether A has determinant 1 or -1.

5.2. Prove that if, in the notation of the previous Exercise, $\det A = 1$, then A is the matrix of the rotation through the angle θ about the origin, counterclockwise.

5.3. Prove that if, in the notation of Exercise 5.1, $\det A = -1$, then A is the matrix of a reflection.

5.4. Check that

$$u = \begin{pmatrix} \cos\phi/2 \\ \sin\phi/2 \end{pmatrix} \quad \text{and} \quad v = \begin{pmatrix} -\sin\phi/2 \\ \cos\phi/2 \end{pmatrix}$$

are eigenvectors with the eigenvalues 1 and -1 for the matrix

$$\begin{pmatrix} \cos\phi & \sin\phi \\ \sin\phi & -\cos\phi \end{pmatrix}.$$

5.5. Use trigonometric identities to prove that

$$\begin{pmatrix} \cos\phi & -\sin\phi \\ \sin\phi & \cos\phi \end{pmatrix} \cdot \begin{pmatrix} \cos\psi & -\sin\psi \\ \sin\psi & \cos\psi \end{pmatrix} = \begin{pmatrix} \cos(\phi+\psi) & -\sin(\phi+\psi) \\ \sin(\phi+\psi) & \cos(\phi+\psi) \end{pmatrix}.$$

Give a geometric interpretation of this fact.

5.6. Prove that any finite group of rotations of the Euclidean plane $\mathbb{R}^2$ about the origin is cyclic.

5.7. Prove that if r is a rotation of $\mathbb{R}^2$ and s a reflection then sr is a reflection; in particular, $|sr| = 2$. Deduce from this the fact that s *inverts* r, i.e., $srs^{-1} = r^{-1}$.

5.8. If G is a finite group of orthogonal transformations of the 2-dimensional Euclidean space $\mathbb{R}^2$ then the map

$$\det : G \longrightarrow \{1, -1\},$$
$$A \mapsto \det A,$$

is a homomorphism with kernel R consisting of all rotations contained in G. If $R \neq G$ then $|G : R| = 2$ and all elements in $G \smallsetminus R$ are reflections.

5.9. Prove that the product of two reflections in $\mathbb{R}^2$ (with a common fixed point at the origin) is a rotation through twice the angle between their mirrors.

INVOLUTARY ORTHOGONAL TRANSFORMATIONS IN THREE DIMENSIONS.

5.10. In $\mathbb{R}^3$ there are three involutive orthogonal transformations, up to conjugacy of matrices, with the eigenvalues $1, 1, -1$ (reflections), $1, -1, -1$, and $-1, -1, -1$. Give a geometric interpretation of the last two transformations.

5.11. An involution in $\mathbb{O}_3$ with the eigenvalues $1, -1, -1$ is called a *half-turn*: it is a rotation through $180°$ around the *axis* spanned by an eigenvector for the eigenvalue 1. Obviously, there is a one-to-one correspondence between one-dimensional subspaces in $\mathbb{R}^3$ and half-turns around them.

1. Prove that the product uvw of three half-turns u, v, w is an involution (and therefore a half-turn) if and only if their axes are coplanar, that is, belong to a 2-dimensional subspace.
2. Furthermore, prove that if $t = uvw$ is an involution then it is a half-turn and its axis is orthogonal to the plane spanned by the axes of $u, v,$ and w.

6

Systems of Mirrors

The philosophy of this book is to use, by all means possible, geometry instead of group theory; this chapter is crucial for the development of our approach. Informally speaking, we introduce systems of mirrors by reflecting a mirror in a mirror, as in a kaleidoscope. We shall soon see that this is a very powerful and useful metaphor.

6.1 Systems of Mirrors

Assume now that we are given a solid $\Delta \subset \mathbb{AR}^n$.

Consider the set Σ of all mirrors of symmetry of Δ, i.e., the mirrors of reflections that send Δ to Δ. The reader can easily check (Exercise 6.2) that Σ is a *closed system of mirrors* in the sense of the following definition: a system of hyperplanes (mirrors) in $\mathbb{AR}^n$ is called *closed* if for any two mirrors H_1 and H_2 in Σ, the mirror image of H_2 in H_1 also belongs to Σ (see Figure 6.1).

Slightly abusing language, we shall call a finite closed system Σ of mirrors simply a *system of mirrors*.

Systems of mirrors are the most natural objects. The reader most likely has seen them when looking into a kaleidoscope; and of course, everybody has seen a mirror. We are interested in the study of finite closed systems of mirrors and other closely related objects—root systems and finite groups generated by reflections.

If Σ is a system of mirrors, the set of all reflections in mirrors of Σ will be referred to as a *closed system of reflections*. In view of Lemma 5.3, a set S of reflections forms a closed system of reflections if and only if $s^t \in S$ for all reflections $s, t \in S$. Here s^t is the standard abbreviation, in group theory, for conjugation:

$$s^t = t^{-1}st.$$

Recall that conjugation by any element t is an automorphism of any group containing t:

$$(xy)^t = x^t y^t.$$

A.V. Borovik and A. Borovik, *Mirrors and Reflections: The Geometry of Finite Reflection Groups*, Universitext, DOI 10.1007/978-0-387-79066-4_6,
© Springer Science+Business Media, LLC 2010

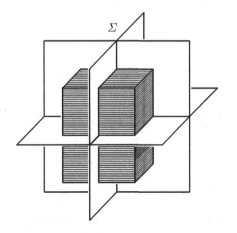

The system Σ of mirrors of symmetry of a geometric body Δ is closed: the reflection of a mirror in another mirror is a mirror again. Notice that if Δ is compact then all mirrors intersect in a common point.

Fig. 6.1. A closed system of mirrors.

Lemma 6.1. *A finite closed system of reflections generates a finite group of isometries.*

Proof. This result is a special case of the following elementary group-theoretic property.

Let W be a group generated by a finite set S of involutions such that $s^t \in S$ for all $s, t \in S$. Then W is finite.

Indeed, since $s \in S$ are involutions, $s^{-1} = s$. Let $w \in W$ and find the shortest expression $w = s_1 \cdots s_k$ of w as a product of elements from S. If the word $s_1 \cdots s_k$ contains two occurrences of the same involution $s \in S$ then

$$
\begin{aligned}
w &= s_1 \cdots s_i s s_{i+1} \cdots s_j s s_{j+1} \cdots s_k \\
&= s_1 \cdots s_i (s_{i+1} \cdots s_j)^s s_{j+1} \cdots s_k \\
&= s_1 \cdots s_i s_{i+1}^s \cdots s_j^s s_{j+1} \cdots s_k \\
&= s_1 \cdots s_i s_{i+1}' \cdots s_j' s_{j+1} \cdots s_k,
\end{aligned}
$$

where all $s_l' = s_l^s$ belong to S and the resulting expression is shorter than the original. Therefore all shortest expressions of elements from W in terms of generators $s \in S$ contain no repetition of symbols. Therefore the length of any such expression is at most $|S|$, and counting the numbers of expressions of length $0, 1, \ldots, |S|$, we find that their total number is at most

$$
1 + |S| + |S|^2 + \cdots + |S|^{|S|}.
$$

Hence W is finite. □

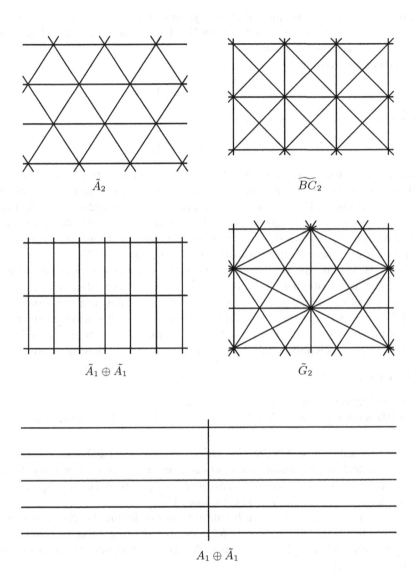

Fig. 6.2. Examples of infinite closed mirror systems in $\mathbb{AR}^2$ with their traditional notation: tesselations of the plane by congruent equilateral triangles ($\tilde{A}_2$), isosceles right triangles ($\widetilde{BC}_2$), rectangles ($\tilde{A}_1 \oplus \tilde{A}_1$), triangles with the angles $\pi/2$, $\pi/3$, $\pi/6$ ($\tilde{G}_2$), infinite half-stripes ($A_1 \oplus \tilde{A}_1$).

6.2 Finite Reflection Groups

A group-theoretic interpretation of closed systems of mirrors comes in the form of a *finite reflection group*, i.e., a finite group W of isometries of an affine Euclidean space A generated by reflections.

Let s be a reflection in W and

$$s^W = \{\, wsw^{-1} \mid w \in W \,\}$$

its conjugacy class. Form the set of mirrors

$$\Sigma = \{\, H_t \mid t \in s^W \,\}.$$

Then it follows from Lemma 5.3 that Σ is a mirror system: if $H_r, H_t \in \Sigma$ then the reflection of H_r in H_t is the mirror H_{r^t}. Thus the conjugacy class s^W is a closed system of reflections. The same observation is valid for any *normal* set S of reflections in W, i.e., a set S such that $s^w \in S$ for all $s \in S$ and $w \in W$. We shall show later that if the reflection group W arises from a closed system of mirrors Σ then every reflection in W is actually the reflection in one of the mirrors in Σ.

Since W is finite, all its orbits are finite and W fixes a point by virtue of Theorem 2.1. We can take this fixed point for the origin of an orthonormal coordinate system and, in view of Theorem 2.2, treat W as a group of linear orthogonal transformations.

If W is the group generated by the reflections in the finite closed system of mirrors Σ then the fixed points of W are fixed by every reflection in a mirror from Σ and hence belong to each mirror in Σ. Thus we have proved the following theorem.

Theorem 6.2.

(1) *A finite reflection group in $\mathbb{AR}^n$ has a fixed point.*
(2) *All the mirrors in a finite closed system of mirrors in $\mathbb{AR}^n$ have a point in common.*

Since we are interested in finite closed systems of mirrors and finite groups generated by reflections, this result allows us to assume without loss of generality that all mirrors pass through the origin of $\mathbb{R}^n$. So we can forget about the affine space $\mathbb{AR}^n$ and work entirely in the Euclidean vector space $V = \mathbb{R}^n$.

Note in passing that there is a beautiful theory of infinite locally finite closed system of mirrors in $\mathbb{AR}^n$ (see Figure 6.2); *locally finite* means that every sphere in $\mathbb{AR}^n$ intersects only finitely many mirrors. A classical exposition of that theory can be found in [Bou].

Exercises

Systems of mirrors.

6.1. We cannot resist temptation and recall an old puzzle: why is it that a mirror changes left and right but does not change up and down?

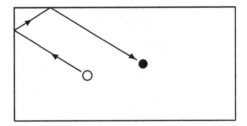

Fig. 6.3. Billiards, for Exercise 6.3.

6.2. Prove that if Δ is a subset in $\mathbb{AR}^n$ then the system Σ of its mirrors of symmetry is closed.

6.3. Two balls, white and black, are placed on a billiard table (Figure 6.3). The white ball must bounce off two cushions of the table and then strike the black one. Find its trajectory.

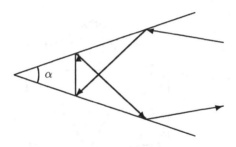

Fig. 6.4. For Exercise 6.4.

6.4. Prove that a ray of light reflecting from two mirrors forming a corner will eventually get out of the corner (Figure 6.4). If the angle formed by the mirrors is α, what is the maximal possible number of times the ray would bounce off the sides of the corner?

6.5. Two big floor-to-ceiling mirrors form the angle α. Inside the angle, a man holds a candle. How many reflections of the candle can the man see?

6.6. Prove that the angular reflector made of three pairwise perpendicular mirrors in $\mathbb{R}^3$ sends a ray of light back in the direction exactly opposite to the one it came from (Figure 6.5).

6.7. How many mirrors of symmetry has a regular tetrahedron? A cube?

6.8. Figure 6.6 shows an icosahedron, one of the five *Platonic solids*. Assuming that it is as symmetric as it appears to be, count the number of its mirrors of symmetry.

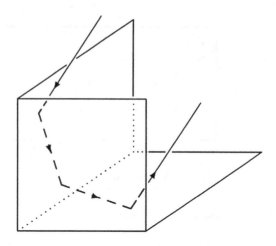

Fig. 6.5. Angular reflector (for Exercise 6.6).

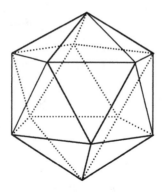

Fig. 6.6. Icosahedron.

REFLECTIONS AND LINEAR ALGEBRA.

6.9. We say that a subspace U of the real Euclidean space V is *perpendicular* to the subspace W and write $U \perp W$ if $U = (U \cap W) \oplus U'$, where U' is orthogonal to W, i.e., $u \cdot w = 0$ for all $u \in U'$ and $w \in W$. Prove that this relation is symmetric: $U \perp W$ if and only if $W \perp U$.

6.10. Prove that if a reflection s leaves a subspace $U < V$ invariant then U is perpendicular to the mirror H_s of the reflection s.

6.11. Prove that two reflections s and t commute, that is, $st = ts$, if and only if their mirrors are perpendicular to each other.

PLANAR GEOMETRY.

6.12. Prove that the product of two reflections in $A\mathbb{R}^2$ with parallel mirrors is a parallel translation. What is the translation vector?

6.13. If a bounded figure in the Euclidean plane $A\mathbb{R}^2$ has a center of symmetry and a mirror of symmetry then it has two perpendicular mirrors of symmetry. Is the same true in $A\mathbb{R}^3$?

7

Dihedral Groups

In this chapter we shall study finite groups generated by two involutions. In subsequent applications, it is usually the group generated by two reflections; the corresponding mirror system is very simple and is of fundamental importance.

7.1 Groups Generated by Two Involutions

Theorem 7.1. *There is a unique, up to isomorphism, group W generated by two involutions s and t such that their product st has order n. Furthermore,*

(1) W *is finite and* $|W| = 2n$.
(2) *If $r = st$ then the cyclic group $R = \langle r \rangle$ generated by r is a normal subgroup of W of index* 2.
(3) *Every element in $W \smallsetminus R$ is an involution.*

We shall denote the group W by Dih_{2n}, call it the *dihedral group* of order $2n$, and write
$$\mathrm{Dih}_{2n} = \langle\, s, t \mid s^2 = t^2 = (st)^n = 1 \,\rangle.$$

This standard group-theoretic notation means that the group Dih_{2n} is generated by two elements s and t such that any identity relating them to each other is a consequence of the *defining relations*

$$s^2 = 1, \ t^2 = 1, \ (st)^n = 1.$$

The words "consequence of the defining relations" are given precise meaning in the theory of groups given by generators and relations, a very well developed chapter of the general theory of groups. We prefer to use them in an informal way that will be always clear from context. The concept of generators and relations will be made more specific in Theorem 15.1. Notice that Theorem 7.1 is a special case of Theorem 15.1.

A.V. Borovik and A. Borovik, *Mirrors and Reflections: The Geometry of Finite Reflection Groups*, Universitext, DOI 10.1007/978-0-387-79066-4_7,
© Springer Science+Business Media, LLC 2010

7.2 Proof of Theorem 7.1

First of all, notice that since $s^2 = t^2 = 1$, we have

$$s^{-1} = s \text{ and } t^{-1} = t.$$

In particular, $(st)^{-1} = t^{-1}s^{-1} = ts$. Set $r = st$. Then

$$r^t = trt = t \cdot st \cdot t = ts = r^{-1}$$

and analogously $r^s = r^{-1}$.

Since $s = rt$, the group W is generated by r and t, and therefore every element w in W has the form

$$w = r^{m_1} t^{k_1} \cdots r^{m_l} t^{k_l},$$

where m_i takes the values $0, 1, \ldots, n - 1$ and k_i is 0 or 1. But one can check that since $trt = r^{-1}$, we have

$$tr = r^{-1}t,$$
$$tr^m = r^{-m}t,$$

and

$$t^k r^m = r^{(-1)^k m} t^k.$$

Hence

$$(r^{m_1} t^{k_1})(r^{m_2} t^{k_2}) = r^m t^k, \tag{7.1}$$

where

$$k = k_1 + k_2, \quad m = m_1 + (-1)^{k_1} m_2.$$

Therefore every element in W can be written in the form

$$w = r^m t^k, \quad m = 0, \ldots, n - 1, \quad k = 0, 1.$$

Furthermore, this presentation is unique. Indeed, assume that

$$r^{m_1} t^{k_1} = r^{m_2} t^{k_2},$$

where $m_1, m_2 \in \{0, \ldots, n - 1\}$ and $k_1, k_2 \in \{0, 1\}$. If $k_1 = k_2$ then $r^{m_1} = r^{m_2}$ and $m_1 = m_2$. But if $k_1 \neq k_2$ then

$$r^{m_1 - m_2} = t.$$

Set $m = m_1 - m_2$. Then $m < n$ and $r^m = (st)^m = t$. If $m = 0$ then $t = 1$, which contradicts our assumption that $|t| = 2$. Now we can easily get a final contradiction:

$$st \cdot st \cdots st = t$$

implies

$$s \cdot ts \cdots ts \cdots ts = 1.$$

The word on the left contains an odd number of elements s and t. Consider the element r in the very center of the word; r is either s or t. Hence the previous equation can be rewritten as

$$sts \cdots r \cdots sts = [sts \cdots] \cdot r \cdot [sts \cdots]^{-1} = 1,$$

which implies $r = 1$, a contradiction.

Since elements of w can be represented by expressions $r^m t^k$, and in a unique way, we conclude that $|W| = 2n$ and

$$W = \{ r^m t^k \mid m = 0, 1, \ldots, n - 1, \ k = 0, 1 \},$$

with the multiplication defined by equation (7.1). This proves existence and uniqueness of Dih_{2n}.

Since $|r| = n$, the subgroup $R = \langle r \rangle$ has index 2 in W and hence is normal in W. If $w \in W \smallsetminus R$ then $w = r^m t$ for some m, and a direct computation shows, that

$$w^2 = r^m t \cdot r^m t = r^{-m+m} t^2 = 1.$$

Since $w \neq 1$, w is an involution. $\qquad\qquad\qquad\qquad\qquad\qquad\qquad\qquad$ □

7.3 Dihedral Groups: Geometric Interpretation

Theorem 7.2. *The group of symmetries* $\mathrm{Sym}\,\Delta$ *of the regular n-gon Δ is isomorphic to the dihedral group* Dih_{2n}.

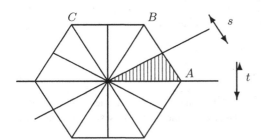

The group of symmetries of the regular n-gon Δ is generated by two reflections s and t in the mirrors passing through the midpoint and a vertex of a side of Δ.

Fig. 7.1. For the proof of Theorem 7.2.

Proof. Set $W = \mathrm{Sym}\,\Delta$. The mirrors of symmetry of the polygon Δ cut it into $2n$ triangular slices (*chambers* in the terminology of Section 3.2); see Figure 7.1. Notice that any two adjacent slices are interchanged by the reflection in their common side. Therefore W acts transitively on the set S of all slices. Also, observe that only the identity symmetry of Δ maps a slice onto itself. By the well-known formula for the length of a group orbit,

$$|W| = \begin{pmatrix} \text{number} \\ \text{of slices} \end{pmatrix} \cdot \begin{pmatrix} \text{number of} \\ \text{elements} \\ \text{fixing a slice} \end{pmatrix} = 2n \cdot 1 = 2n.$$

Next, if s and t are reflections in the side mirrors of a slice, then their product st is a rotation through the angle $2\pi/n$, which can be immediately seen from the picture: st maps the vertex A to B and B to C. (Notice that we use the "left" notation for action, so when we apply the composition st of two transformations s and t to a point, we apply t first and then s: $(st)A = s(tA)$.) By Theorem 7.1, $|\langle s, t \rangle| = 2n$; hence $W = \langle s, t \rangle$ is the dihedral group of order $2n$.   □

The system of mirrors of symmetry of the regular n-gon is traditionally denoted by $G_2(n)$ (some books use different, no less traditional notation: $I_2(n)$). It contains n mirrors; the reflections in the mirrors make n involutions in the reflection group Dih_{2n}.

Exercises

7.1. Prove that the dihedral group Dih_6 is isomorphic to the symmetric group Sym_3.

7.2. THE CENTER OF A DIHEDRAL GROUP. If $n > 2$ then

$$Z(\mathrm{Dih}_{2n}) = \begin{cases} \{\, 1 \,\} & \text{if } n \text{ is odd,} \\ \{\, 1, r^{\frac{n}{2}} \,\} = \langle r^{\frac{n}{2}} \rangle & \text{if } n \text{ is even.} \end{cases}$$

Here, $Z(X)$ is the standard notation for the *center* of the group X, that is, the set of elements in X that commute with every element in X.

7.3. KLEIN FOUR-GROUP. Prove that Dih_4 is an abelian group,

$$\mathrm{Dih}_4 = \{\, 1, s, t, st \,\}.$$

(It is traditionally called the *Klein Four-Group*.)

7.4. Prove that the dihedral group Dih_{2n}, $n > 2$, has one class of conjugate involutions, if n is odd, and three classes, if n is even. In the latter case, one of the classes contains just one involution z and $Z(\mathrm{Dih}_{2n}) = \{\, 1, z \,\}$.

7.5. Prove that there exists a group, unique up to isomorphism, generated by two involutions such that their product has infinite order. (The group is called the *infinite dihedral group*.)

7.6. Prove that a finite group of orthogonal transformations of $\mathbb{R}^2$ is either cyclic or a dihedral group Dih_{2n}.

7.7. If $W = \mathrm{Dih}_{2n}$ is a dihedral group of orthogonal transformations of $\mathbb{R}^2$, then W has one conjugacy class of reflections if n is odd, and two conjugacy classes of reflections if n is even.

7.8. Check that the complex numbers

$$e^{2k\pi i/n} = \cos 2k\pi/n + i \sin 2k\pi/n, \quad k = 0, 1, \ldots, n-1,$$

in the complex plane $\mathbb{C}$ are vertices of a regular n-gon Δ. Prove that the maps

$$r : z \mapsto z \cdot e^{2\pi i/n},$$
$$t : z \mapsto \bar{z},$$

where $^-$ denotes complex conjugation, generate the group of symmetries of Δ.

7.9. Use the idea of the proof of Theorem 7.2 to find the orders of the groups of symmetries of the regular tetrahedron, cube, dodecahedron.

8

Root Systems

In this chapter we introduce objects that are in a sense "dual" to mirror systems: given a mirror system, we take, for each mirror, a pair of normal vectors. If lengths of these normal vectors are chosen in a coordinated way, the resulting system of vectors is preserved by all reflections; in that case we say that we have a *root system*.

Root systems provide a more traditional approach to finite reflection groups—and have exceptionally important applications on their own. In our approach, we freely use both mirrors and roots with the aim of maximizing the intuitive geometric aspect of the theory.

8.1 Mirrors and Their Normal Vectors

Consider a reflection s with the mirror H. If we choose the orthogonal system of coordinates in V with the origin O belonging to H, then s fixes O and thus can be treated as a linear orthogonal transformation of V. Let us take a nonzero vector α perpendicular to H. Then obviously, $\mathbb{R}\alpha = H^{\perp}$ is the orthogonal complement of H in V, and s preserves $H^{\perp}$ and therefore sends α to $-\alpha$. Then we can easily check that s can be written in the form

$$s_{\alpha}\beta = \beta - \frac{2\beta \cdot \alpha}{\alpha \cdot \alpha}\alpha,$$

where $\alpha \cdot \beta$ denotes the scalar product of α and β. Indeed, a direct computation shows that the formula holds when $\beta \in H$ and when $\beta = \alpha$. By the obvious linearity of the right side of the formula with respect to β, it is also true for all $\beta \in H + \mathbb{R}\alpha = V$.

Also we can check by a direct computation (left to the reader as an exercise) that given the nonzero vector α, the linear transformation s_{α} is orthogonal, i.e.,

$$s_{\alpha}\beta \cdot s_{\alpha}\gamma = \beta \cdot \gamma$$

for all vectors β and γ. Finally,

$$s_{\alpha} = s_{c\alpha}$$

A.V. Borovik and A. Borovik, *Mirrors and Reflections: The Geometry of Finite Reflection Groups*, Universitext, DOI 10.1007/978-0-387-79066-4_8,
© Springer Science+Business Media, LLC 2010

for any nonzero scalar c.

We know that reflections can be characterized as linear orthogonal transformations of $\mathbb{R}^n$ with one eigenvalue -1 and $(n-1)$ eigenvalues 1 (Lemma 5.2); the vector α in this case is an eigenvector corresponding to the eigenvalue -1.

Thus we have a one-to-one correspondence between the following three classes of objects:

- hyperplanes (i.e., vector subspaces of codimension 1) in the Euclidean vector space V;
- nonzero vectors defined up to multiplication by a nonzero scalar;
- reflections in the group of orthogonal transformations of V.

The mirror H of the reflection s_α will be denoted by H_α. Notice that $H_\alpha = H_{c\alpha}$ for any nonzero scalar c.

Notice, finally, that orthogonal linear transformations of the Euclidean vector space V (with the origin O fixed) preserve the relations between mirrors, vectors, and reflections.

8.2 Root Systems

Traditionally, closed systems of reflections were studied in the disguise of *root systems*. By definition, a finite set Φ of vectors in V is called a root system if it satisfies the following two conditions:

(1) $\Phi \cap \mathbb{R}\rho = \{\rho, -\rho\}$ for all $\rho \in \Phi$;
(2) $s_\rho \Phi = \Phi$ for all $\rho \in \Phi$.

The following lemma is an immediate corollary of Lemma 5.3.

Lemma 8.1. *Let Σ be a finite closed system of mirrors. For every mirror H in Σ take two vectors $\pm\rho$ of length 1 perpendicular to H. Then the collection Φ of all these vectors is a root system. Conversely, if Φ is a root system then $\{\, H_\rho \mid \rho \in \Phi \,\}$ is a system of mirrors.*

Proof. We need only recall that a reflection s, being an orthogonal transformation, preserves orthogonality of vectors and hyperplanes: if ρ is a vector and H is a hyperplane then $\rho \perp H$ if and only if $s\rho \perp sH$. $\qquad\square$

Notice that if Φ is a root system then the vectors $\rho/|\rho|$ with $\rho \in \Phi$ form the root system Φ' with the same mirror system and the same reflection group (Figure 8.1). In this book, we are not much interested in lengths of roots and in most cases can assume that all roots have length 1. However, in vast areas of application of reflection groups it makes sense to work with root systems with different root lengths; see, for example, the beautiful root system in Figure 8.4.

We can also restate Lemma 6.1 in terms of root systems.

Lemma 8.2. *Let Φ be a root system. Then the group W generated by reflections s_ρ for $\rho \in \Phi$ is finite.*

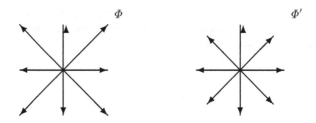

Fig. 8.1. If Φ is a root system then the vectors $\rho/|\rho|$ with $\rho \in \Phi$ form the root system Φ' with the same reflection group. We are not much interested in lengths of roots and in most cases can assume that all roots have length 1.

8.3 Planar Root Systems

We wish to begin the development of the theory of root systems by appealing to the reader's geometric intuition.

Lemma 8.3. *If Φ is a root system in $\mathbb{R}^2$ then the angles formed by pairs of neighboring roots are all equal. (See Figure 8.2.)*

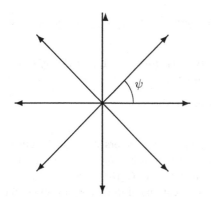

The fundamental property of planar root systems: the angles ψ formed by pairs of neighboring roots are all equal. If the root system contains $2n$ vectors then $\psi = \pi/n$ and the reflection group is the dihedral group Dih_{2n} of order $2n$.

Fig. 8.2. A planar root system (Lemma 8.3).

Proof. The proof of this simple result becomes self-evident if we consider, instead of roots, the corresponding system Σ of mirrors; see Figure 8.3. The mirrors in Σ cut the plane into corners (we call them *chambers*), and adjacent corners, with the angles ϕ and ψ, are congruent because they are mirror images of each other. Therefore all corners are congruent. But the angle between neighboring mirrors is exactly the angle between the corresponding roots. $\square$

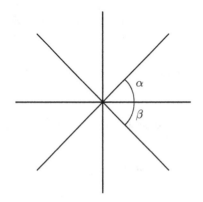

The fact that the angles formed by pairs of neighboring roots are all equal becomes obvious if we consider the corresponding system of mirrors: $\alpha = \beta$ because the adjacent angles are mirror images of each other.

Fig. 8.3. A planar mirror system (for the proof of Lemma 8.3).

Corollary 8.4. *In a planar mirror system, the angles between neighboring mirrors are all equal.*

Lemma 8.5. *If a planar root system Φ contains $2n$ vectors, $n \geqslant 1$, then the reflection group $W(\Phi)$ is the dihedral group Dih_{2n} of order $2n$.*

Proof. The proof is left to the reader as an exercise. $\square$

We see that a planar root system consisting of $2n$ vectors of equal length is uniquely defined, up to elation of $\mathbb{R}^2$. We shall denote it by $G_2(n)$. Later we shall introduced planar root systems A_2 (which coincides with $G_2(3)$) as a part of series of n-dimensional root systems A_n. In many applications of the theory of reflection groups the lengths of roots are of importance; in particular, the root system $G_2(4)$ associated with the system of mirrors of symmetry of the square comes in two versions, named B_2 and C_2, which contain eight roots of two different lengths; see Figure 9.5 in Section 9.2. Finally, the regular hexagon gives rise to the root system of type $G_2(6)$ (which is traditionally notated just G_2); see Figure 8.4.

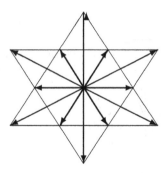

Fig. 8.4. The root system G_2.

8.4 Positive and Simple Systems

Let $f : \mathbb{R}^n \longrightarrow \mathbb{R}$ be a linear function. Assume that f does not vanish on roots in Φ, i.e., $f(\alpha) \neq 0$ for all $\alpha \in \Phi$. Then every root ρ in Φ is called *positive* or *negative* according to whether $f(\rho) > 0$ or $f(\rho) < 0$. We shall write, abusing notation, $\alpha > \beta$ if $f(\alpha) > f(\beta)$. The system of all positive roots will be denoted by Φ^+ and called the *positive system*. Correspondingly, the *negative system* is denoted by Φ^-. Obviously

$$\Phi = \Phi^+ \sqcup \Phi^-.$$

Let Γ denote the convex polyhedral cone spanned by the positive system Φ^+. We follow the notation of Section 4 and call the positive roots directed along the edges of Γ *simple* roots. The set of all simple roots is called the *simple system* of roots and denoted by Π; roots in Π are called *simple* roots. It is intuitively evident that the cone Γ is generated by simple roots; see also Lemma 4.3. In particular, every root ϕ in Φ^+ can be written as a nonnegative combination of roots in Π:

$$\phi = c_1 \rho_1 + \cdots + c_m \rho_m, \quad c_i \geqslant 0, \quad \rho_i \in \Pi.$$

Notice that the definition of positive, negative, and simple systems depends on the choice of the linear function f. We shall call a set of roots positive, negative, or simple if it is so for some linear function f.

Lemma 8.6. *In a simple system Π, the angle between two distinct roots is nonacute:*

$$\alpha \cdot \beta \leqslant 0$$

for all $\alpha \neq \beta$ in Π.

Proof. Let P be a two-dimensional plane spanned by α and β. Define $\Phi_0 = \Phi \cap P$. If $\gamma, \delta \in \Phi_0$ then the reflection s_γ maps δ to the vector

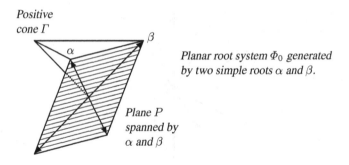

Positive
cone Γ

Planar root system Φ_0 generated
by two simple roots α and β.

Plane P
spanned by
α and β

Fig. 8.5. For the proof of Lemma 8.6

$$s_\gamma \delta = \delta - \frac{2\gamma \cdot \delta}{\gamma \cdot \gamma} \gamma,$$

which obviously belongs to P and Φ_0. Hence every reflection s_γ for $\gamma \in \Phi_0$ obviously maps P to P and Φ_0 to Φ_0. This means that Φ_0 is a root system in P and $\Phi^+ \cap P$ is a positive system in Φ_0.

Moreover, the convex polyhedral cone Γ_0 spanned by $\Phi_0^+ = \Phi^+ \cap P$ is contained in $\Gamma \cap P$. Since α and β are obviously directed along the edges of $\Gamma \cap P$ (see also Lemma 4.4) and belong to Γ_0, it follows that $\Gamma_0 = \Gamma \cap P$ and α and β belong to a simple system in Φ_0; see Figure 8.5. Therefore the lemma is reduced to the 2-dimensional case, where it is self-evident; see Figure 8.6.   □

Notice that our proof of Lemma 8.6 is a manifestation of a general principle: surprisingly many considerations in root systems can be reduced to computations with pairs of roots.

Our proof of the following result, Theorem 8.7, follows [Hum, Theorem 1.3], but deviates from the latter by emphasizing the reduction to the two-dimensional case.

Theorem 8.7. *Every simple system Π is linearly independent. In particular, every root β in Φ can be written, and in a unique way, in the form $\sum c_\alpha \alpha$, where $\alpha \in \Pi$ and all coefficients c_α are either nonnegative (when $\beta \in \Phi^+$) or nonpositive (when $\beta \in \Phi^-$).*

Proof. Assume, by way of contradiction, that Π is linearly dependent and

$$\sum_{\alpha \in \Pi} a_\alpha \alpha = 0,$$

where some coefficient a_α is nonzero. Separate positive and negative coefficients a_α and rewrite this equality as

$$\sum b_\beta \beta = \sum c_\gamma \gamma,$$

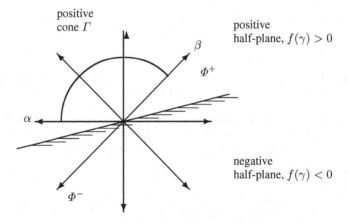

Fig. 8.6. For the proof of Lemma 8.6. In the 2-dimensional case the nonacuteness of the simple system is obvious: the roots α and β are directed along the edges of the convex cone spanned by Φ^+, and the angle between α and β is at least $\pi/2$.

where the coefficients b_β and c_γ are strictly positive and the sums are taken over disjoint subsets of Π. Set

$$\sigma = \sum b_\beta \beta.$$

Since all roots β are positive, $\sigma \neq 0$. But

$$0 \leqslant \sigma \cdot \sigma = \sum_\beta b_\beta \beta \cdot \sum_\gamma c_\gamma \gamma = \sum_\beta \sum_\gamma b_\beta c_\gamma \beta \cdot \gamma \leqslant 0,$$

because all individual scalar products $\beta \cdot \gamma$ are nonpositive by Lemma 8.6. Therefore $\sigma = 0$, a contradiction. $\qquad\square$

Corollary 8.8. *All simple systems in Φ contain an equal number of roots.*

Proof. Indeed, it follows from Theorem 8.7 that a simple system is a maximal linearly independent subset of Φ. $\qquad\square$

The number of roots in a simple system of the root system Φ is called the *rank* of Φ and denoted by $\mathrm{rk}\,\Phi$. The subscript n in the standard notation for root systems A_n, B_n, etc. (which will be introduced later) refers to their ranks.

Exercises

8.1. Prove, by direct computation, that the linear transformation s_α given by the formula

$$s_\alpha \beta = \beta - \frac{2\beta \cdot \alpha}{\alpha \cdot \alpha} \alpha$$

is orthogonal, that is,

$$s_\alpha \beta \cdot s_\alpha \gamma = \beta \cdot \gamma$$

for all $\beta, \gamma \in V$.

8.2. Let Φ be a root system in the Euclidean space V and $U < V$ a vector subspace of V. Prove that $\Phi \cap U$ is a (possibly empty) root system in U.

8.3. Let V_1 and V_2 be two subspaces orthogonal to each other in the real Euclidean vector space V and let Φ_i be a root system in V_i, $i = 1, 2$. Prove that $\Phi = \Phi_1 \cup \Phi_2$ is a root system in $V_1 \oplus V_2$; it is called the *direct sum* of Φ_1 and Φ_2 and denoted by

$$\Phi = \Phi_1 \oplus \Phi_2.$$

8.4. We say that a group W of orthogonal transformations of V is *essential* if it acts on V without nonzero fixed points. Let Φ be a root system in V, Φ and W the corresponding systems of mirrors and reflection groups. Prove that the following conditions are equivalent:

- Φ spans V.
- The intersection of all mirrors in Σ consists of one point.
- W is essential on V.

8.5. Prove Lemma 8.5.

8.6. Prove that in a root system in $\mathbb{R}^2$, the lengths of roots can take at most two values.

8.7. In a root system of a two-dimensional reflection group Dih_{2n} with odd n, all roots have the same length.

8.8. Describe planar root systems with two and four roots and the corresponding reflection groups.

8.9. Use the observation that the root system G_2 contains two subsystems of type A_2 to show that the dihedral group Dih_{12} contains two different subgroups isomorphic to the dihedral group Dih_6.

8.10. Prove that in a planar root system $\Phi \subset \mathbb{R}^2$, all positive systems are conjugate under the action of the reflection group $W = W(\Phi)$. Prove the same for simple systems.

9

Root Systems A_{n-1}, BC_n, D_n

This chapter contains an ad hoc construction of the most important classes of mirror and root systems, together with a detailed discussion of their properties. The reader will encounter many familiar geometric objects.

9.1 Root System A_{n-1}

The root system A_{n-1} is arguably the most important and ubiquitous. This is not surprising; for as we shall soon see, it is intimately related to the mother of all finite groups, the symmetric group Sym_n of all permutations of the set $\{1, 2, \ldots, n\}$.

9.1.1 A Few Words about Permutations

We shall use both "two row" and cycle notation for permutations. For example, if

$$w = \begin{pmatrix} 1\ 2\ 3\ 4\ 5 \\ 2\ 1\ 4\ 5\ 3 \end{pmatrix},$$

then the same permutation can also be written as

$$w = (12)(345);$$

in both cases it is the map

$$w : 1 \mapsto 2,\ 2 \mapsto 1,\ 3 \mapsto 4,\ 4 \mapsto 3,\ 5 \mapsto 3.$$

We will write the action of permutations on the left, so that

$$w \cdot 1 = 2,\ w \cdot 3 = 4.$$

Notie that this means that st is the permutation t followed by s if s and t are permutations. Perhaps some readers are used to the opposite convention.

A.V. Borovik and A. Borovik, *Mirrors and Reflections: The Geometry of Finite Reflection Groups*, Universitext, DOI 10.1007/978-0-387-79066-4_9, © Springer Science+Business Media, LLC 2010

9.1.2 Permutation Representation of Sym_n

Let V be the real vector space $\mathbb{R}^n$ with the standard orthonormal basis $\epsilon_1, \ldots, \epsilon_n$ and the corresponding coordinates $x_1, \ldots, x_n$.

The group $W = \mathrm{Sym}_n$ acts on V in the natural way, by permuting the n vectors $\epsilon_1, \ldots, \epsilon_n$:

$$w\epsilon_i = \epsilon_{wi},$$

which obviously induces an action of W on Φ. The action of the group $W = \mathrm{Sym}_n$ on $V = \mathbb{R}^n$ preserves the standard scalar product associated with the orthonormal basis $\epsilon_1, \ldots, \epsilon_n$. Therefore W acts on V by orthogonal transformations.

In its action on V the *transposition* $r = (ij)$ acts as the reflection in the mirror of symmetry given by the equation $x_i = x_j$.

Lemma 9.1. *Every reflection in W is a transposition.*

Proof. The cycle $(i_1 \ldots i_k)$ has exactly one eigenvalue 1 when restricted to the subspace

$$\mathbb{R}\epsilon_{i_1} \oplus \cdots \oplus \mathbb{R}\epsilon_{i_k},$$

with the eigenvector

$$\epsilon_{i_1} + \cdots + \epsilon_{i_k}.$$

It follows from this observation that the multiplicity of the eigenvalue 1 of the permutation $w \in \mathrm{Sym}_n$ equals the number of cycles in the cycle decomposition of w (we have also to count the trivial one-element cycles of the form (i)). If w is a reflection, then the number of cycles is $n - 1$; hence w is a transposition. $\qquad\square$

9.1.3 Regular Simplices

The convex hull Δ of the points $\epsilon_1, \ldots, \epsilon_n$ is the convex polytope defined by the equation and inequalities

$$x_1 + \cdots + x_n = 1, \ x_1 \geqslant 0, \ldots, x_n \geqslant 0.$$

Since the group $W = \mathrm{Sym}_n$ permutes the vertices of Δ, it acts as a group of symmetries of Δ. We shall denote the full group of symmetries of Δ by $\mathrm{Sym}\,\Delta$, so that $W \leqslant \mathrm{Sym}\,\Delta$. We wish to prove that actually $W = \mathrm{Sym}\,\Delta$. Indeed, any symmetry s of Δ acts on the set of vertices as some permutation $w \in \mathrm{Sym}_n$; hence the symmetry $s^{-1}w$ fixes all the vertices $\epsilon_1, \ldots, \epsilon_n$ of Δ and therefore is the identity symmetry.

The polytope Δ is called the *regular $(n-1)$-simplex*. When $n = 3$, Δ is an equilateral triangle lying in the plane $x_1 + x_2 + x_3 = 1$ (see Figure 9.1), and when $n = 4$, Δ is a regular tetrahedron lying in the 3-dimensional affine Euclidean space

$$x_1 + x_2 + x_3 + x_4 = 1.$$

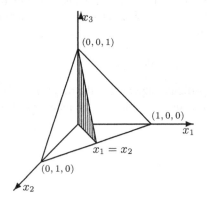

The transposition (12) *acts on* $\mathbb{R}^3$ *as the reflection in the mirror* $x_1 = x_2$ *and as a symmetry of the equilateral triangle with the vertices*

$$(1,0,0),\ (0,1,0),\ (0,0,1).$$

Fig. 9.1. Sym_n is the group of symmetries of the regular simplex

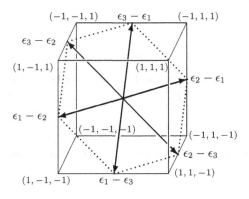

The root system

$$\{\, \epsilon_i - \epsilon_j \mid i \neq j \,\}$$

of type A_2 lies in the hyperplane

$$x_1 + x_2 + x_3 = 0,$$

which cuts a regular hexagon in the unit cube $[-1,1]^3$.

Fig. 9.2. Root system of type A_2

9.1.4 The Root System A_{n-1}

We shall introduce the root system Φ of type A_{n-1} as the system of vectors in $V = \mathbb{R}^n$ of the form $\epsilon_i - \epsilon_j$, where $i, j = 1, 2, \ldots, n$ and $i \neq j$ (Figure 9.2). Notice that Φ is invariant under the action of $W = \mathrm{Sym}_n$ on V.

In its action on V the transposition $r = (ij)$ acts as the reflection in the mirror of symmetry perpendicular to the root $\rho = \epsilon_i - \epsilon_j$. Hence Φ is a root system. Since the symmetric group is generated by transpositions, $W = W(\Phi)$ is the corresponding reflection group, and the mirror system Σ consists of all hyperplanes $x_i = x_j$, $i \neq j$, $i, j = 1, \ldots, n$.

Notice that the group W is not essential for V; indeed, it fixes all points in the 1-dimensional subspace $\mathbb{R}(\epsilon_1 + \cdots + \epsilon_n)$ and leaves invariant the $(n-1)$-dimensional linear subspace U defined by the equation

$$x_1 + \cdots + x_n = 0.$$

It is easy to see that $\Phi \subset U$ spans U. In particular, the rank of the root system Φ is n, which justifies the use, in accordance with our convention, of the index $n - 1$ in the notation A_{n-1} for it.

9.1.5 The Standard Simple System

Take the linear function

$$f(x) = x_1 + 2x_2 + \cdots + nx_n.$$

Obviously f does not vanish on roots, and the corresponding positive system has the form

$$\Phi^+ = \{\, \epsilon_i - \epsilon_j \mid j < i \,\}.$$

The set of positive roots

$$\Pi = \{\, \epsilon_2 - \epsilon_1, \epsilon_3 - \epsilon_2, \ldots, \epsilon_n - \epsilon_{n-1} \,\}$$

is linearly independent, and every positive root is obviously a linear combination of roots in Π with nonnegative coefficients: for $i > j$,

$$\epsilon_i - \epsilon_j = (\epsilon_i - \epsilon_{i-1}) + \cdots + (\epsilon_{j+1} - \epsilon_j).$$

Therefore Π is a simple system. It is called the *standard simple system* of the root system A_{n-1}.

9.1.6 Action of Sym_n on the Set of all Simple Systems

The following result is a partial case of Theorem 11.6. But the elementary proof given here is instructive on its own.

Lemma 9.2. *The group* $W = \mathrm{Sym}_n$ *acts simply transitively on the set of all positive (respectively simple) systems in* Φ.

Proof. Since there is a natural one-to-one correspondence between simple and positive systems, it is enough to prove that W acts simply transitively on the set of positive systems in Φ.

Let f be an arbitrary linear function that does not vanish on Φ, that is, $f(\epsilon_i - \epsilon_j) \neq 0$ for all $i \neq j$. Then all the values

$$f(\epsilon_1), \ldots, f(\epsilon_n)$$

are different, and we can list them in strictly increasing order:

$$f(\epsilon_{i_1}) < f(\epsilon_{i_2}) < \cdots < f(\epsilon_{i_n}).$$

Now consider the permutation w given, in column notation, as

$$w = \begin{pmatrix} 1 & 2 & \cdots & n-1 & n \\ i_1 & i_2 & \cdots & i_{n-1} & i_n \end{pmatrix}.$$

Thus the function f defines a new ordering, which we shall denote by $\leqslant^w$, on the set $[n]$:

$$j \leqslant^w i \text{ if and only if } f(\epsilon_j) \leqslant f(\epsilon_i).$$

If we look again at the table for w, we see that above any element i in the bottom row lies the element $w^{-1}i$ in the upper row. Thus

$$i \leqslant^w j \text{ if and only if } w^{-1}i \leqslant w^{-1}j.$$

Notice also that the permutation w and the associated ordering $\leqslant^w$ of $[n]$ uniquely determine each other.[1]

Now consider the positive system Φ_0^+ defined by the linear function f,

$$\Phi_0^+ = \{\, \epsilon_i - \epsilon_j \mid f(\epsilon_i - \epsilon_j) > 0 \,\}.$$

We have the following chain of equivalences:

$$\begin{aligned}
\epsilon_i - \epsilon_j \in \Phi_0^+ \text{ iff } & f(\epsilon_j) < f(\epsilon_i) \\
\text{iff } & j <^w i \\
\text{iff } & w^{-1}j < w^{-1}i \\
\text{iff } & \epsilon_{w^{-1}i} - \epsilon_{w^{-1}j} \in \Phi^+ \\
\text{iff } & w^{-1}(\epsilon_i - \epsilon_j) \in \Phi^+ \\
\text{iff } & \epsilon_i - \epsilon_j \in w\Phi^+.
\end{aligned}$$

This proves that $\Phi_0^+ = w\Phi^+$ and also that the permutation w is uniquely determined by the positive system Φ_0^+. Since Φ^+, by its construction from an arbitrary linear function, represents an arbitrary positive system in Φ, the group W acts on the set of positive systems in Φ simply transitively.   □

---

[1] In the nineteenth century the orderings, or rearrangements, of $[n]$ were called *permutations*, and the permutations in the modern sense, i.e., maps from $[n]$ to $[n]$, were called *substitutions*. These are two aspects, "passive" and "active," of the same object. We shall see later that they correspond to treating a permutation as an element of a reflection group Sym_n or an element of the Coxeter complex for Sym_n.

9.2 Root Systems of Types C_n and B_n

9.2.1 Hyperoctahedral Group

The hyperoctahedral group is the group of symmetries of the n-cube $[-1,1]^n$ (Figure 9.3) in the n-dimensional real Euclidean space $\mathbb{R}^n$. However, it will be convenient to describe it first in purely combinatorial terms.

Let

$$[n] = \{1, 2, \ldots, n\} \text{ and } [n]^* = \{1^*, 2^*, \ldots, n^*\}.$$

Define the map

$$* : [n] \longrightarrow [n]^*$$

by

$$i \mapsto i^*$$

and the map

$$* : [n]^* \longrightarrow [n]$$

by

$$(i^*)^* = i.$$

Then $*$ is an involutive permutation[2] of the set $[n] \sqcup [n]^*$.

Let W be the group of all permutations of the set $[n] \sqcup [n]^*$ that commute with the involution $*$, i.e., a permutation w belongs to W if and only if

$$w(i^*) = w(i)^*$$

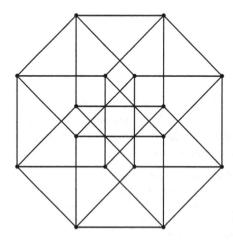

Fig. 9.3. A 2-dimensional projection of the 4-dimensional cube

---

[2] That is, a permutation of order 2.

for all $i \in [n] \sqcup [n]^*$. We shall call permutations with this property *admissible*. The group W is known as the *hyperoctahedral group BC_n*. It is easy to see that W is indeed isomorphic to the group of symmetries of the n-cube $[-1,1]^n$ in the n-dimensional real Euclidean space $\mathbb{R}^n$. Indeed, if $\epsilon_1, \epsilon_2, \ldots, \epsilon_n$ is the standard orthonormal basis in $\mathbb{R}^n$, we set, for $i \in [n]$,

$$\epsilon_{i^*} = -\epsilon_i.$$

Then we can define the action of W on $\mathbb{R}^n$ by the following rule:

$$w\epsilon_i = \epsilon_{wi}.$$

Since w is an admissible permutation of $[n] \sqcup [n]^*$, the linear transformation is well defined and orthogonal. Also it can be easily seen that W is exactly the group of all orthogonal transformations of $\mathbb{R}^n$ preserving the set of vectors $\{\pm\epsilon_1, \pm\epsilon_2, \ldots, \pm\epsilon_n\}$ and thus preserving the cube $[-1,1]^n$. Indeed, the vectors $\pm\epsilon_i$, $i \in [n]$, are exactly the unit vectors normal to the $(n-1)$-dimensional faces of the cube (given, obviously, by the linear equations $x_i = \pm 1$, $i = 1, 2, \ldots, n$).

The name "hyperoctahedral" for the group W is justified by the fact that the group of symmetries of the n-cube coincides with the group of symmetries of its dual polytope, whose vertices are the centers of the faces of the cube. The dual polytope for the n-cube is known as the *n-cross polytope* or *n-dimensional hyperoctahedron*; see Figure 9.4.

9.2.2 Admissible Orderings

We shall order the set $[n] \sqcup [n]^*$ in the following way:

$$n^* < n-1^* < \cdots < 2^* < 1^* < 1 < 2 < \cdots < n-1 < n.$$

If now $w \in W$, then we define a new ordering $\leqslant^w$ of the set $[n] \sqcup [n]^*$ by the rule

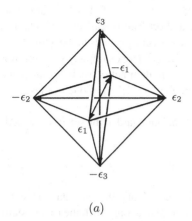 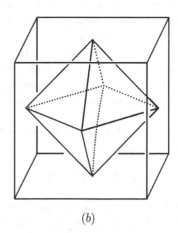

$$(a) \qquad\qquad\qquad\qquad (b)$$

Fig. 9.4. Hyperoctahedron ("octahedron" in dimension $n = 3$), or n-cross polytope, is the convex hull of the points $\pm\epsilon_i$, $i = 1, \ldots, n$, in $\mathbb{R}^n$ (picture (a)). Obviously the hyperoctahedron is the dual polytope to the unit cube (picture (b)).

$$i \leqslant^w j \quad \text{if and only if} \quad w^{-1}i \leqslant w^{-1}j.$$

Orderings of the form $\leqslant^w$, $w \in W$, are called *admissible* orderings of the set $[n] \sqcup [n]^*$. They can be characterized by the following property:

an ordering $\prec$ on $[n] \sqcup [n]^*$ is admissible if and only if from $i \prec j$ it follows that $j^* \prec i^*$.

Conversely, if $\prec$ is an admissible ordering, then the permutation

$$w = \begin{pmatrix} n^* & (n-1)^* & \ldots & 1^* & 1 & \ldots & n-1 & n \\ j_1 & j_2 & & \cdots & j_n & j_{n+1} & \cdots & j_{2n-1} & j_{2n} \end{pmatrix},$$

where

$$j_1 \prec j_2 \prec \cdots \prec j_{2n-1} \prec j_{2n},$$

is admissible and the ordering $\prec$ coincides with $\leqslant^w$.

9.2.3 Root Systems C_n and B_n

Let ϵ_i, $i \in [n]$, be the standard orthonormal basis in $\mathbb{R}^n$, and set

$$\epsilon_{i^*} = -\epsilon_i$$

for $i^* \in [n]^*$. This defines the vectors ϵ_j for all $j \in [n] \sqcup [n]^*$. Now the root system Φ of type C_n is formed by the vectors $2\epsilon_j$, $j \in [n] \sqcup [n]^*$ (called *long roots*), together with the vectors $\epsilon_{j_1} - \epsilon_{j_2}$, where $j_1, j_2 \in [n] \sqcup [n]^*$, $j_1 \neq j_2$ or j_2^* (called *short roots*). Written in the standard basis $\epsilon_1, \epsilon_2, \ldots, \epsilon_n$, the roots take the form $\pm 2\epsilon_i$ or $\pm \epsilon_i \pm \epsilon_j$, $i, j = 1, 2, \ldots, n$, $i \neq j$. Notice that both short and long roots can be written as $\epsilon_j - \epsilon_i$ for some $i, j \in [n] \sqcup [n]^*$.

It is easy to see that when ρ is one of the long roots $\pm 2\epsilon_i$, $i \in [n]$, then s_ρ is the linear transformation corresponding to the element (i, i^*) of W in its canonical representation. Analogously, if $\rho = \epsilon_i - \epsilon_j$ is a short root (recall that we use the convention $\epsilon_{i^*} = -\epsilon_i$ for $i \in [n]$), then the reflection s_ρ corresponds to the admissible permutation $(i, j)(i^*, j^*)$. Moreover, one can easily check (for example, by computing the eigenvalues of admissible permutations from W in their action on $\mathbb{R}^n$) that every reflection in the group of symmetries of the unit cube $[-1, 1]^n$ is of one of these two types.

Now we see that the use of the name "root system" in regard to the set Φ is justified. The root system

$$B_n = \{ \pm \epsilon_i \pm \epsilon_j, \pm \epsilon_i \mid i, j = 1, 2, \ldots, n, i \neq j \}$$

differs from C_n only in lengths of roots (see Figures 9.5 and 9.6) and has the same reflection group BC_n. Therefore in the sequel we shall deal only with the root system C_n.

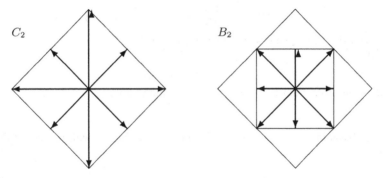

C_2 B_2

Fig. 9.5. Root systems B_2 and C_2.

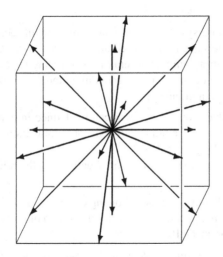

Fig. 9.6. The root system B_3 inscribed in the unit cube $[-1, 1]^3$.

9.2.4 Action of W on Φ

Now consider the linear functional

$$f(x) = x_1 + 2x_2 + 3x_3 + \cdots + nx_n.$$

It is easy to see that a root $\epsilon_i - \epsilon_j$ is positive with respect to f if in the ordering

$$n^* < n - 1^* < \cdots < 1^* < 1 < 2 < \cdots < n$$

of the set $[n] \sqcup [n]^*$, we have $i > j$. The system of positive roots Φ^+ associated with f is called the *standard positive system of roots*. The set

$$\Pi = \{\, 2\epsilon_1, \epsilon_2 - \epsilon_1, \ldots, \epsilon_n - \epsilon_{n-1} \,\}$$

is obviously the simple system of roots contained in Φ^+.

If now

$$j_1 <^w j_2 <^w \cdots <^w j_{2n-1} <^w j_{2n}$$

is an admissible ordering of $[n] \sqcup [n]^*$, then the vectors

$$\epsilon_{j_{n+1}}, \epsilon_{j_{n+2}}, \ldots, \epsilon_{j_{2n}}$$

form a basis in $\mathbb{R}^n$. Let $y_1, y_2, \ldots, y_n$ be the coordinates with respect to this basis and

$$f(y) = y_1 + 2y_2 + 3y_3 + \cdots + n y_n.$$

Then obviously, f does not vanish on roots in Φ, and for a root $\epsilon_j - \epsilon_i$ in Φ, the inequality $f(\epsilon_j - \epsilon_i) > 0$ is equivalent to $i \leqslant^w j$. Thus the system of positive roots associated with f coincides with the system

$$w\Phi^+ = \{\epsilon_j - \epsilon_i \mid i \leqslant^w j\}$$

obtained from the standard system Φ^+ of positive roots by the action of the element w. Obviously, the simple system of roots contained in Φ^+ is exactly $w\Pi$.

If now Π' is an arbitrary simple system of roots arising from an arbitrary linear function $f : \mathbb{R}^n \longrightarrow \mathbb{R}$ not vanishing on roots in Φ, then the following objects are uniquely determined by our choice of Π':

- the system of positive roots $\Phi^{+\prime}$, which can be defined in two equivalent ways: as the set of all roots that are nonnegative linear combinations of roots from Π', and as the set $\{r \in \Phi \mid f(r) > 0\}$;
- the (obviously admissible) ordering $\prec$ on J defined by the rule $i \prec j$ if and only if $f(\epsilon_i) \leqslant f(\epsilon_j)$.

In particular, we immediately have the following observation (which is a partial case of a more general result about conjugacy of simple system of roots for arbitrary finite reflection groups, Theorem 11.10).

Proposition 9.3. *Any two simple systems in the root system Φ of type B_n or C_n are conjugate under the action of W. Moreover, the reflection group W is simply transitive in its action on the set of simple systems in Φ.*

9.3 The Root System D_n

By definition,

$$D_n = \{\pm\epsilon_i \pm \epsilon_j \mid i, j = 1, 2, \ldots, n, i \neq j \,\};$$

thus D_n is a subsystem of the root system C_n.

The system

$$\Pi = \{\, \epsilon_1 + \epsilon_2, \epsilon_2 - \epsilon_1, \epsilon_3 - \epsilon_2, \ldots, \epsilon_n - \epsilon_{n-1} \,\}$$

is a simple system in Φ.

A fact worth special mentioning is that the root system D_3 coincides with the root system A_3; see Figure 9.7.

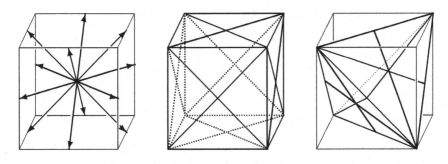

Fig. 9.7. Shown are the root system D_3 inscribed in the unit cube $[-1, 1]^3$ (on the left) and the corresponding mirror system (shown in the middle by intersections with the surface of the cube and the tetrahedron inscribed in the cube). Comparing the last two pictures, we see that the mirror system of type D_3 is isometric to the mirror system of type A_3.

Exercises

Fig. 9.8. 2-dimensional projection of a 4-dimensional cross polytope.

9.1. Can you convince yourself that Figure 9.8 indeed represents a 2-dimensional projection of a 4-dimensional cross polytope?

9.2. Make a sketch of the root systems $A_1 \oplus A_1$ in $\mathbb{R}^2$ and $A_1 \oplus A_1 \oplus A_1$ in $\mathbb{R}^3$.

9.3. Check that when we take the intersections of the mirrors of reflections in $W = \mathrm{Sym}_n$ to the subspace $x_1 + \cdots + x_n = 0$ of $\mathbb{R}^n$, the resulting system of mirrors can be geometrically described as the system of mirrors of symmetry of the regular $(n-1)$-simplex with the vertices

$$\delta_i = \epsilon_i - \frac{1}{n}(\epsilon_1 + \cdots + \epsilon_n), \quad i = 1, \ldots, n.$$

9.4. An orthogonal transformation of the Euclidean space $\mathbb{R}^n$ is called a *rotation* if its determinant is 1. For a polytope Δ, denote by $\mathrm{Rot}\,\Delta$ the subgroup of $\mathrm{Sym}\,\Delta$ formed by all rotations. We know that the group of symmetries $\mathrm{Sym}\,\Delta$ of the regular tetrahedron Δ in $\mathbb{R}^3$ is isomorphic to Sym_4; prove that $\mathrm{Rot}\,\Delta$ is the alternating group Alt_4.

9.5. Prove that every reflection in BC_n has the form (ii^*) or $(ij)(i^*j^*)$.

9.6. Prove that the reflection group of type BC_2 is isomorphic to the dihedral group Dih_8.

9.7. THE GROUP OF SYMMETRIES OF THE CUBE. Observe that the group $W = BC_3$ of symmetries of the cube $\Delta = [-1, 1]^3$ contains the involution z that sends every vertex of the cube to its opposite.

1. Check that $\det z = -1$, so that z does not belong to the group $R = \mathrm{Rot}\,\Delta$ of rotations of the cube.
2. Prove that $z \in Z(W)$. (Here, $Z(X)$ is the standard notation for the *center* of the group X, that is, the set of elements in X that commute with every element in X.)
3. Prove that the group R acts faithfully on the set D of 4 diagonals of the cube Δ, that is, the segments connecting the opposite vertices of the cube. Moreover, every permutation of diagonals is the result of the action of a rotation of the cube. Hence $R \simeq \mathrm{Sym}_4$.
4. Prove that $W = \langle z \rangle \times R$.
5. Prove that $\langle z \rangle = Z(W)$.
6. Prove that the symmetries of the cube that send every 2-dimensional face of the cube into itself or the opposite face form a normal abelian subgroup $E < W$ of order 8. Prove further that $W/E \simeq \mathrm{Sym}_3$ and that actually $W = E \rtimes T$ for some subgroup $T \simeq \mathrm{Sym}_3$.

9.8. IMPORTANT ROOT SUBSYSTEMS.
Prove that
1. the set Θ of roots $\{ \pm\epsilon_i \mid i = 1, \ldots, n \}$ is a root system of type $A_1 \oplus \cdots \oplus A_1$ (n summands);
2. the intersection Ψ of Φ with the hyperplane $x_1 + \cdots + x_n = 0$ is a root system of type A_{n-1}.

9.9. Prove that the group of symmetries of the n-cube $[-1, 1]^n$ is indeed BC_n.

9.10. (R. Sandling) Prove that

$$\mathrm{Sym}\,[-1, 1]^n = \{ w \in \mathrm{GL}_n(\mathbb{R}^n) \mid w([-1, 1]^n) = [-1, 1]^n \},$$

i.e., linear transformations preserving the cube are in fact orthogonal.

9.11. THE STRUCTURE OF THE HYPEROCTAHEDRAL GROUP. Use Exercise 9.8 to show that if E and T are the reflection groups corresponding to the systems of roots Θ and Ψ then

1. $E \simeq \mathbb{Z}_2 \times \cdots \times \mathbb{Z}_2$ (n factors);
2. $E \lhd W$;
3. $T \simeq \mathrm{Sym}_n$;
4. $W = E \rtimes T$.

9.12. The standard cross polytope in $\mathbb{R}^n$ is the convex hull of the $2n$ points

$$(\pm 1, 0, \ldots, 0), (0, \pm 1, 0, \ldots, 0), \ldots, (0, \ldots, 0, \pm 1).$$

Let Δ be its orthogonal projection on the hyperplane

$$x_1 + \cdots + x_n = 1.$$

Find $\mathrm{Sym}(\Delta)$. Identify (that is, find the name of) Δ when $n = 2, 3, 4$.

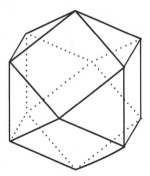 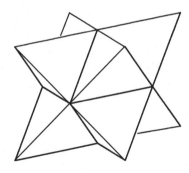

Fig. 9.9. Cuboctahedron (left) and stella octangula (right).

9.13. Describe explicitly an isometry between the root systems

$$D_3 = \{ \pm\epsilon_i \pm \epsilon_j \mid i, j = 1, 2, 3, i \neq j \}$$

and

$$A_3 = \{ \epsilon_i - \epsilon_j \mid i, j = 1, 2, 3, 4, i \neq j \}$$

(see Figure 9.7).

9.14. Sketch the root system D_2; you will see that it consists of two orthogonal pairs of vectors, each forming the 1-dimensional system A_1. Thus $D_2 = A_1 \oplus A_1$.

9.15. Find the groups of symmetries of the cubooctahedron and stella octangula (Figure 9.9). Are they BC_3 or D_3?

9.16. Find a natural homomorhism from the group BC_3 of symmetries of the cube onto the symmetric group Sym_4. What is the kernel of this homomorphism?

Part III

Coxeter Complexes

10

Chambers

In this chapter we finally start a systematic development of the general theory of reflection groups.

Consider the system Σ of all mirrors of reflections s_ρ for $\rho \in \Phi$. Of course, this is a hyperplane arrangement in the sense of Chapter 3, and we shall freely use the relevant terminology. In particular, chambers of Σ are open polyhedral cones—connected components of

$$V \smallsetminus \bigcup_{H \in \Sigma} H.$$

The closures of these cones are called *closed chambers*. Facets of chambers (i.e., faces of maximal dimension) are *panels* and have mirrors in Σ as *walls*. Notice that every panel belongs to a unique wall. To fully appreciate this architectural terminology (introduced by J. Tits), imagine a building built out of walls of double-sided mirrors. Two chambers are *adjacent* if they have a panel in common. Notice that every chamber is adjacent to itself.

Theorem 10.1. *Every chamber C has the form*

$$C = \bigcap_{\rho \in \Pi} V_\rho^-$$

for some simple system Π. Every panel of C belongs to one of the walls H_ρ for a root $\rho \in \Pi$. Conversely, if $\rho \in \Pi$ then $H_\rho \cap \overline{C}$ is a panel of C.

Proof. Take any vector γ in the chamber C and consider the linear function

$$f(\lambda) = -\gamma \cdot \lambda.$$

Since γ does not belong to any mirror H_α in Σ, the function f does not vanish on roots in Φ. Therefore the condition $f(\alpha) > 0$ determines a positive system Φ^+ and the corresponding simple system Π. Now consider the cone C' defined by the

A.V. Borovik and A. Borovik, *Mirrors and Reflections: The Geometry of Finite Reflection Groups*, Universitext, DOI 10.1007/978-0-387-79066-4_10,

Positive cone Γ

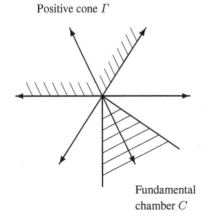

The fundamental chamber C is defined as the interior of the cone dual to the positive cone Γ, i.e., as the set of vectors λ such that $\lambda \cdot \gamma < 0$ for all $\gamma \in \Gamma$.

Fundamental
chamber C

Fig. 10.1. The fundamental chamber.

inequalities $(\lambda, \rho) < 0$ for all $\rho \in \Pi$. Obviously $\gamma \in C'$ and therefore $C \subseteq C'$. If $C \neq C'$ then some hyperplane H_α, $\alpha \in \Phi$, bounds C and intersects C' nontrivially. But $\alpha = \sum c_\rho \rho$, where all c_ρ are all nonnegative or all nonpositive, and

$$\gamma \cdot \alpha = \sum c_\rho \gamma \cdot \rho$$

cannot be equal to 0. This contradiction shows that $C = C'$. The closure $\overline{C}$ of C is defined by the inequalities $\lambda \cdot \rho \leqslant 0$ for $\rho \in \Pi$, which is equivalent to $\lambda \cdot \alpha \leqslant 0$ for $\alpha \in \Gamma$. Therefore the cone $\overline{C}$ is dual to the positive cone Γ (see Figure 10.1), and every facet of C is perpendicular to some edge of Γ, and vice versa. In particular, every panel of C belongs to the wall H_ρ for some simple root $\rho \in \Pi$.

The same argument works in the reverse direction: if Π is any simple system, then since Π is linearly independent, we can find a vector γ such that $\gamma \cdot \rho < 0$ for all $\rho \in \Pi$. Then $\gamma \cdot \alpha \neq 0$ for all roots $\alpha \in \Phi$, and the chamber C containing γ has the form

$$C = \bigcap_{\rho \in \Pi} V_\rho^-.$$

$\square$

If Π is a simple system then the corresponding chamber is called the *fundamental chamber* of C (Figure 10.1).

The set of all chambers associated with the root system Φ is called the *Coxeter complex* and will be denoted by $\mathcal{C}$. See, for example, Figures 10.2, 10.3 and 10.4.

The following lemma is an immediate consequence of Lemma 3.3.

Lemma 10.2. *The union of two distinct adjacent closed chambers is convex.*

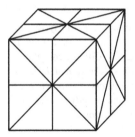

The Coxeter complex of type BC_3 is formed by all the mirrors of symmetry of the cube; here they are shown by their lines of intersection with the faces of the cube.

Fig. 10.2. The Coxeter complex BC_3.

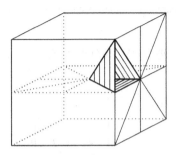

Fig. 10.3. A chamber in the Coxeter complex BC_3.

Exercises

10.1. Show that there is a one-to-one correspondence between the chambers of the mirror system A_{n-1} and the elements in the symmetric group $W = \mathrm{Sym}_n$: associate with the permutation

$$\begin{pmatrix} 1 & 2 & \ldots & n-1 & n \\ i_1 & i_2 & \ldots & i_{n-1} & i_n \end{pmatrix}$$

the open cone in the hyperplane $x_1 + \cdots + x_n = 0$ given in $\mathbb{R}^n$ by the system of inequalities

$$x_{i_1} < x_{i_2} < \cdots < x_{i_n}.$$

Prove that this cone is a chamber of the mirror system associated with the standard root system for Sym_n.

10.2. Analogously, prove that the set of all admissible orderings on the set $[n] \sqcup [n]^*$ can be put in one-to-one correspondence with the set of chambers for the group $W = BC_n$ if we associate with each admissible ordering

$$i_n^* < \cdots < i_1^* < i_1 < \cdots < i_n$$

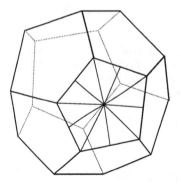

In the case that Σ is the system of mirrors of symmetry of a regular polytope Δ, the Coxeter complex is basically the subdivision of the faces of Δ by the mirrors of symmetries of faces (here shown only on one face of the dodecahedron Δ).

Fig. 10.4. Chambers for the system of mirrors of a regular polytope.

the open cone in $\mathbb{R}^n$ given by the system of inequalities

$$0 < \tilde{x}_{i_1} < \cdots < \tilde{x}_{i_n};$$

we use here the convention that

$$\tilde{x}_i = \begin{cases} x_i & \text{if } i \in [n], \\ -x_{i^*} & \text{if } i \in [n]^*. \end{cases}$$

11

Generation

It is time to build closer links between geometry (mirror systems) and group theory (the corresponding reflection group).

11.1 Simple Reflections

Let $\Pi = \{ \rho_1, \ldots, \rho_n \}$ be a simple system of roots. The corresponding reflections $r_i = s_{\rho_i}$ are called *simple reflections*. (Figure 11.1.)

The symmetry group of the tetrahedron acts on its four vertices as the symmetric group Sym_4. *The reflections in the walls of the fundamental chamber are the transpositions* (12), (23), *and* (34). *Therefore they generate* Sym_4.

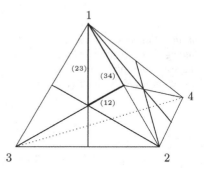

Fig. 11.1. Generation by simple reflections (Theorem 11.1).

Theorem 11.1. *The group* W *is generated by a simple system of reflections.*

Proof. Set $W' = \langle r_1, \ldots, r_n \rangle$. We shall prove first that

the group W' *is transitive in its action on* $\mathcal{C}$.

Proof of the claim. The fundamental chamber C is bounded by panels lying on the mirrors of the simple reflections $r_1, \ldots, r_n$. Therefore the neighboring chambers (i.e.,

A.V. Borovik and A. Borovik, *Mirrors and Reflections: The Geometry of Finite Reflection Groups*, Universitext, DOI 10.1007/978-0-387-79066-4_11,
© Springer Science+Business Media, LLC 2010

the chambers sharing a common mirror with C) can be obtained from C by reflections in these mirrors; they are $r_1 C, \ldots, r_n C$. Now let $w \in W'$. Then the panels of the chamber wC belong to the mirrors of reflections $w r_1 w^{-1}, \ldots, w r_n w^{-1}$. If D is a chamber adjacent to wC then it can be obtained from wC by reflecting wC in the common mirror; hence $D = w r_i w^{-1} \cdot wC = w r_i C$ for some $i = 1, \ldots, n$. Notice that $w r_i \in W'$. We can proceed to move from a chamber to an adjacent one until we present all chambers in $\mathcal{C}$ in the form wC for appropriate elements $w \in W$.   □

We can now complete the proof. If $\alpha \in \Phi$ is any root and s_α the corresponding reflection, then the wall H_α bounds some chamber D. We know that $D = wC$ for some $w \in W'$. The fundamental chamber C is bounded by the walls H_{ρ_i} for simple roots ρ_i (Theorem 10.1), and therefore the wall H_α equals $w H_{\rho_i}$ for some simple root ρ_i. Thus $s_\alpha = w r_i w^{-1}$ belongs to W'. Since the group W is generated by reflections s_α we have $W = W'$.   □

In the course of the proof we have obtained one more important result:

Corollary 11.2. *The action of W on $\mathcal{C}$ is transitive.*

This observation will be later incorporated into Theorem 11.6.

11.2 Foldings

In the 2-dimensional case, a folding is exactly what its name suggests: the plane is being folded on itself like a sheet of paper.

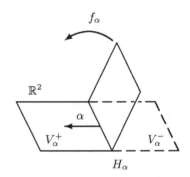

Fig. 11.2. Folding.

Given a nonzero vector $\alpha \in V$, the hyperplane

$$H_\alpha = \{ \gamma \in V \mid \gamma \cdot \alpha = 0 \}$$

cuts V into two subspaces

$$V_\alpha^+ = \{ \gamma \mid \gamma \cdot \alpha \geqslant 0 \} \text{ and } V_\alpha^- = \{ \gamma \mid \gamma \cdot \alpha \leqslant 0 \}$$

intersecting along the common hyperplane H_α. The *folding* in the direction of α is the map f_α defined by the formula

$$f_\alpha(\beta) = \begin{cases} \beta & \text{if } \beta \cdot \alpha \geqslant 0, \\ s_\alpha\beta & \text{if } \beta \cdot \alpha < 0. \end{cases}$$

Thus f_α fixes all points in V_α^+ and maps V_α^- onto V_α^+ symmetrically (see Figure 11.2). Notice that f_α is an idempotent map, i.e., $f_\alpha f_\alpha = f_\alpha$. The folding $f_{-\alpha}$ is called the *opposite* to f_α. The reflection s_α is made up of two foldings f_α and $f_{-\alpha}$:

$$s_\alpha = f_\alpha|_{V_\alpha^+} \cup f_{-\alpha}|_{V_\alpha^-} .$$

We say that a folding f *covers* a subset $X \subset V$ if $X \subseteq f(V)$.

By definition, a folding of the chamber complex $\mathcal{C}$ is a folding along one of its walls.

Proposition 11.3. *A folding f of $\mathcal{C}$ sends chambers to chambers and preserves adjacency: if C and D are two adjacent chambers then their images $f(C)$ and $f(D)$ are also adjacent. (Remember that by definition of adjacency, this includes the possibility that $f(C) = f(D)$.)*

11.3 Galleries and Paths

Given two chambers C and D, we can always find a sequence G of chambers

$$C = C_0, C_1, \ldots, C_{l-1}, C_l = D$$

such that every two consecutive chambers C_{i-1} and C_i are adjacent. We shall call G a *gallery* connecting the chambers C and D. Notice that our definition of adjacency allows two adjacent chambers to coincide. This means that we also allow repetition of chambers in a gallery: it could happen that $C_{i-1} = C_i$. We shall say in this situation that the gallery *stutters* at chamber C_i. The number l will be called the *length* of the gallery G.

Notice that if s_i is the reflection in a common wall of two adjacent chambers C_{i-1} and C_i, then either $C_i = s_i C_{i-1}$ or $C_i = C_{i-1}$.

Given $w \in W$, we wish to describe a canonical way of connecting the fundamental chamber C and the chamber $D = wC$ by a gallery. We know that W is generated by the fundamental reflections $r_1, \ldots, r_n$, i.e., the reflections in the walls of the fundamental chamber C. The minimal number l such that w is the product of some l fundamental reflections is called the *length* of w and denoted by $l(w)$.

Let $w = r_{i_1} \cdots r_{i_l}$. We leave to the reader to check the following group-theoretic identity: since all r_i are involutions,

$$r_{i_1} \cdots r_{i_l} = r_{i_l}^{r_{i_{l-1}} \cdots r_{i_1}} \cdot r_{i_{l-1}}^{r_{i_{l-2}} \cdots r_{i_1}} \cdots r_{i_2}^{r_{i_1}} \cdot r_{i_1}.$$

Define

$$s_j = r_{i_j}^{r_{j-1} \cdots r_{i_1}}.$$

Then $w = s_l \cdots s_1$, and moreover,

$$s_j \cdots s_1 = r_{i_1} \cdots r_{i_j} \text{ for } j = 1, \ldots, l.$$

Define by induction $C_0 = C$ and for $i = 1, \ldots, l$, $C_j = s_j C_{i-1}$, so that

$$C_j = s_j \cdots s_1 C_0 = r_{i_1} \cdots r_{i_j} C_0 \text{ for } j > 0,$$

$$C_l = r_{i_1} \cdots r_{i_l} C_0 = wC = D.$$

Notice that $s_1 = r_1$ is the reflection in the common wall of the chambers C_0 and C_1. Next, s_j for $j > 1$ is written as

$$s_j = r_{i_j}^{r_{i_{j-1}} \cdots r_{i_1}} = (r_{i_1} \cdots r_{i_{j-1}}) r_{i_j} (r_{i_1} \cdots r_{i_{j-1}})^{-1}.$$

By Lemma 5.3, since r_{i_j} is a reflection in a panel, say H, of the fundamental chamber $C = C_0$, s_j is the reflection in the panel $r_{i_1} \cdots r_{i_{j-1}} H$ of the chamber $r_{i_1} \cdots r_{i_{j-1}} C_0 = C_{j-1}$. Since $s_j C_{j-1} = C_j$,

s_j *is the reflection in the common panel of the chambers* C_{j-1} *and* C_j.

Summarizing this procedure, we obtain the following result; it will show us the correct path through the labyrinth of mirrors.

Theorem 11.4. *Let* $w = r_{i_1} \cdots r_{i_l}$ *be an expression of* $w \in W$ *in terms of simple reflections* r_i. *Let* C *be the fundamental chamber and* D *a chamber in* C *such that* $D = wC$. *Then there exists a unique gallery* $C_0, C_1, \ldots, C_l$ *connecting* $C = C_0$ *and* $D = C_l$ *with the following property:*

$$s_j = r_{i_j}^{r_{j-1} \cdots r_{i_1}}$$

is the reflection in the common wall of C_{j-1} *and* C_j *for* $j = 1, \ldots, l$, *and*

$$w = s_l \cdots s_1.$$

The gallery $C_0, \ldots, C_l$ constructed in Theorem 11.4 will be called the *canonical* w-*gallery* starting at $C = C_0$.

We can reverse the above arguments and obtain also the following result.

Theorem 11.5. *Let* $C_0, \ldots, C_l$ *be a gallery connecting the fundamental chamber* $C = C_0$ *and a chamber* $D = C_l$. *Assume that the gallery does not stutter at any chamber, that is, no two successive chambers* C_i *and* C_{i+1} *coincide. Let* s_i *be the reflection in the common wall of* C_{i-1} *and* C_i, $i = 1, \ldots, l$.

Then $D = wC$ *for* $w = s_l \cdots s_1$,

$$C_j = s_j \cdots s_1 C_0 \text{ for all } j = 1, \ldots, l,$$

and there exists an expression $w = r_{i_1} \cdots r_{i_l}$ *of* w *in terms of simple reflections* r_i *such that for all* $j = 1, \ldots, l$,

$$s_j = r_{i_j}^{r_{j-1} \cdots r_{i_1}}.$$

11.4 Action of W on $\mathcal{C}$

In this section we shall prove arguably the most important property of the Coxeter complex.

Theorem 11.6. *The group W is simply transitive on $\mathcal{C}$, i.e., for any two chambers C and D in $\mathcal{C}$ there exists a unique element $w \in W$ such that $D = wC$.*

Paths.

We shall call a sequence of points $\gamma_0, \ldots, \gamma_l$ a *path* if

- the consecutive points γ_{i-1} and γ_i are contained in adjacent chambers C_{i-1} and C_i;
- if $C_{i-1} = C_i$ then $\gamma_{i-1} = \gamma_i$;
- if $C_{i-1} \neq C_i$ and s_i is the reflection in the common panel of C_{i-1} and C_i then $\gamma_i = s_i \gamma_{i-1}$.

The number l is called the *length* of the path. Set $w = s_l \cdots s_0$. Since

$$\gamma_l = s_l \cdots s_1 \gamma_0 = w\gamma_0,$$

the sequence of chambers $C_0, C_1, \ldots, C_l$ is the canonical w-gallery, and by Theorem 11.5, w can be expressed as a product of l simple reflections. So we have the following useful lemma.

Lemma 11.7. *Given a path $\gamma_0, \gamma_1, \ldots, \gamma_l$, there exists $w \in W$ such that $\gamma_l = w\gamma_0$ and $l(w) \leqslant l$.*

Notice the important property of paths: since we know that the union of two distinct adjacent closed chambers is convex (Lemma 10.2), the wall H_{s_i} is the only wall intersecting the segment $[\gamma_{i-1}\gamma_i]$. Therefore the following lemma holds.

Lemma 11.8. *If $\gamma_0, \ldots, \gamma_l$ is a path connecting the points γ_0 and γ_l lying on opposite sides of the wall H, then the path intersects H in the sense that for some two consecutive points γ_{i-1} and γ_i, the wall H intersects the segment $[\gamma_{i-1}, \gamma_i]$ and*

- *the common panel of the chambers C_{i-1} and C_i containing γ_{i-1} and γ_i, respectively, belongs to H;*
- *γ_{i-1} and γ_i are symmetric in H.*

11.5 Paths and Foldings

As often happens in the theory of reflection groups, an important technical result we wish to state now can be best justified by referring to a picture (Figure 11.3).

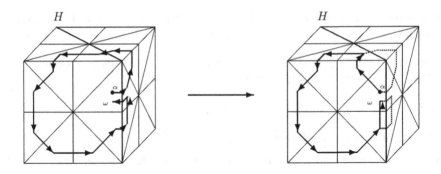

Fig. 11.3. For the proof of Lemma 11.9: the folding in a wall that intersects a path converts the path to a shorter one.

Lemma 11.9. *Assume that the starting point* $\alpha = \gamma_0$ *and the end point* $\omega = \gamma_l$ *of a path* $\gamma_0, \ldots, \gamma_l$ *lie on one side of a wall* H. *If the wall* H *intersects the path, that is, one of the points* γ_i *lies on the opposite side of* H *from* α, *then the path can be replaced by a shorter path with the same starting and end points, and such that it does not intersect the wall* H.

Proof. See the quite self-explanatory Figure 11.3. A rigorous proof follows. However, it can be skipped on first reading.

Assume that the path intersects the wall H at the segment $[\gamma_{p_1-1}\gamma_{p_1}]$. Then, in view of Lemma 11.8, the path should intersect the wall at least once more, say at the segments

$$[\gamma_{p_2-1}\gamma_{p_2}], \ldots, [\gamma_{p_k-1}\gamma_{p_k}].$$

Let $C_0, \ldots, C_l$ be the gallery corresponding to our path, so that $\gamma_i \in C_i$. Take the folding f in H onto the half-space containing α and β and consider the path $f(\gamma_0), f(\gamma_1), \ldots, f(\gamma_l)$ and the gallery $f(C_0), f(C_1), \ldots, f(C_l)$. In this new gallery and new path we have repeated chambers, namely

$$f(C_{p_1-1}) = f(C_{p_1}), \ldots, f(C_{p_k-1}) = f(C_{p_k}),$$

and points

$$f(\gamma_{p_1-1}) = f(\gamma_{p_1}), \ldots, f(\gamma_{p_k-1}) = f(\gamma_{p_k}).$$

After deleting the duplicate chambers and points and changing the numeration we obtain a *shorter* gallery $C'_0, C'_1, \ldots, C'_m$ and a path $\gamma'_0, \gamma'_1, \ldots, \gamma'_m$ such that $\gamma'_0 = \gamma_0$, $\gamma'_m = \gamma_l$ and for all $i = 1, \ldots, m$,

- $\gamma'_i \in C'_i$;
- C'_{i-1} and C_i are adjacent;
- if s'_i is the reflection in the common wall of C'_{i-1} and C'_i then $\gamma'_i = s_i\gamma'_{i-1}$.

But then $= \gamma'_0, \ldots, \gamma'_m$ is a shorter path connecting α and ω. □

11.6 Simple Transitivity of W on $\mathcal{C}$: Proof of Theorem 11.6

In view of Corollary 11.2, we need to prove only the uniqueness of w. If $D = w_1 C$ and $D = w_2 C$ for two elements $w_1, w_2 \in W$ and $w_1 \neq w_2$, then $w_2^{-1} w_1 C = C$. Set $w = w_2^{-1} w_1$; we wish to prove $w = 1$. Assume, by way of contradiction, that $w \neq 1$. Of all expressions of w in terms of the generators $r_1, \ldots, r_n$ we take a shortest, $w = r_{i_1} \cdots r_{i_l}$, where $l = l(w)$ is the length of w. Since $w \neq 1$, $l \neq 0$. Now of all $w \in W$ with the property that $wC \neq C$ choose the one with smallest length l.

We can assume without loss of generality that C is the fundamental chamber. Let now $C_0, C_1, \ldots, C_l$ be the canonical w-gallery connecting C with C.

The vectors from the open cone C obviously span the vector space V, so the nontrivial linear transformation w cannot fix them all. Take $\gamma \in C$ such that $w\gamma \neq \gamma$ and consider the sequence of points γ_i, $i = 0, 1, \ldots, l$, defined by $\gamma_0 = \gamma$ and $\gamma_i = s_i \gamma_{i+1}$ for $i > 0$. Then $\gamma_i \in C_i$. The sequence $\gamma_0, \gamma_1, \ldots, \gamma_l$ is a path and links the endpoints $\gamma_0 = \gamma$ and $\gamma_l = w\gamma$. Now consider the wall $H = H_{s_1}$. Since γ_0 and γ_l both lie in C, they lie on the same side of H. But the point $\gamma_1 = s_1 \gamma_0$ lies on the opposite side of H from γ. Hence, by Lemma 11.9, there is a shorter path connecting γ and $w\gamma$ and, by Lemma 11.7, an element $w' \in W$ with $w'\alpha = \omega$ and smaller length $l(w') < l$ than that of w. This contradiction completes the proof of the theorem. $\quad\square$

Since we have a one-to-one correspondence between positive systems, simple systems, and fundamental chambers, we arrive at the following result.

Theorem 11.10. *The group W acts simply transitively on the set of all positive (simple) systems in Φ.*

Another important result is the following observation: for every root $\alpha \in \Phi$ the mirror H_α bounds one of the chambers in $\mathcal{C}$. Since every chamber corresponds to some simple system in Φ and all simple systems are conjugate by Theorem 11.10, we arrive to the following result.

Theorem 11.11. *Let Φ be a root system, Π a simple system in Φ, and W the reflection group of Φ. Every root $\alpha \in \Phi$ is conjugate, under the action of W, to a root in Π.*

Exercises

11.1. Use Theorem 11.1 to prove the (well-known) fact that the symmetric group Sym_n is generated by transpositions

$$(12), (23), \ldots, (n-1, n)$$

(see Figure 11.1).

11.2. Prove that the reflections

$$r_1 = (12)(1^*2^*), \ldots, r_{n-1} = (n-1, n)(n-1^*, n^*), \; r_n = (n, n^*)$$

generate the hyperoctahedral group BC_n.

11.3. Prove that the reflections

$$r_1 = (12)(1^*2^*), \ldots, r_{n-1} = (n-1, n)(n-1^*, n^*), r_n = (12^*)(1^*2)$$

generate the reflection group D_n viewed as a subgroup of the hyperoctahedral group BC_n.

11.4.* Let T be an arbitrary set of transpositions in the symmetric group Sym_n. We shall associate with T the graph $\Gamma = \Gamma(T)$ constructed as follows. Vertices of Γ are elements in $[n]$, and two vertices a and b are connected by a (nonoriented) edge if and only if the transposition (a, b) belongs to T.

Prove that the set of transpositions T generate the symmetric group Sym_n if and only if the graph Γ is connected.

11.5.* Formulate and prove, by analogy with Exercise 11.4, a reasonably simple criterion for a set of permutations of the form $(ij)(i^*j^*)$, $i, j \in [n] \sqcup [n]^*$, to generate the group D_n.

11.6. When you fold a sheet of paper, why is the line along which it is folded straight?

11.7. There are three foldings of the chamber complex BC_2 such that their composition maps the chamber complex onto one of its chambers. What is the minimal number of foldings needed for folding the chamber complex BC_3 onto one chamber?

11.8. Prove, for involutions $r_1, \ldots, r_l$ in a group G, the identity

$$r_1 \cdots r_l = r_l^{r_{l-1} \cdots r_1} \cdot r_{l-1}^{r_{l-2} \cdots r_1} \cdots r_2^{r_1} \cdot r_1.$$

12

Coxeter Complex

12.1 Labeling of the Coxeter Complex

We shall use the simple transitivity of the action of the reflection group W on its Coxeter complex $\mathcal{C}$ to label each panel of the Coxeter complex $\mathcal{C}$ with one of the simple reflections $r_1, \ldots, r_n$; the procedure for labeling is as follows.

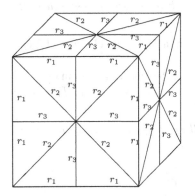 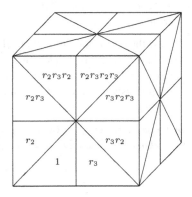

Fig. 12.1. Labeling of panels and chambers in the Coxeter complex C_3.

First we label the panels of the fundamental chamber C by the corresponding simple reflections (see Figure 12.1). If D is a chamber in $\mathcal{C}$, then there is a unique element $w \in W$ that sends C to $D = wC$. If Q is a panel of D, we assign to the panel Q of D the same label as that of the panel $P = w^{-1}Q$ of C.

A.V. Borovik and A. Borovik, *Mirrors and Reflections: The Geometry of Finite Reflection Groups*, Universitext, DOI 10.1007/978-0-387-79066-4_12,
© Springer Science+Business Media, LLC 2010

However, we need to take care of consistency of labeling: the panel Q belongs to two adjacent chambers D and D'. If we label the panels of D' by the same rule, will the label assigned to Q be the same? Let r be the simple reflection in the panel P and $C' = rC$ the chamber adjacent to C and sharing the panel P with C. Since the action of W on $\mathcal{C}$ preserves adjacency of chambers, $D' = wC' = wrC$. Hence wr is the unique element of W that sends C to D', and we assign to the panel Q the label of the panel $(wr)^{-1}Q$ of C. But $rP = P$; hence $(wr)^{-1}Qrw^{-1}Q = rP = P$, and Q gets the same label as before.

If a common panel of two chambers D and E is labeled r_i, we shall say that D and E are r_i-adjacent. This includes the case $D = E$, so that every chamber is r_i-adjacent to itself.

The following observation is immediate.

Proposition 12.1. *The action of W preserves the labeling of panels in the Coxeter complex $\mathcal{C}$.*

Moreover, we can now start to develop a vocabulary for translation of the geometric properties of the Coxeter complex $\mathcal{C}$ into the language of the corresponding reflection group W.

Theorem 12.2. *Let C be a fundamental chamber in the Coxeter complex $\mathcal{C}$ of a reflection group W. The map*

$$w \mapsto wC$$

is a one-to-one correspondence between the elements in W and chambers in $\mathcal{C}$. Two distinct chambers C and C' are r_i-adjacent if and only if the corresponding elements w and w' are related as $w' = wr_i$.

Now the description of canonical galleries given in Theorems 11.4 and 11.5 can be put in a much more convenient form.

Let $\Gamma = \{C_0, \ldots, C_l\}$ be a gallery and let r_{i_k} be the label of the common panel of the successive chambers C_{k-1} and C_k, $k = 1, \ldots, l$. Then we say that Γ has *type* $r_{i_1}, \ldots, r_{i_l}$.

Theorem 12.3. *Let $\Gamma = \{C_0, \ldots, C_l\}$ be a gallery of type $r_{i_1}, \ldots, r_{i_l}$ connecting the fundamental chamber $C = C_0$ and a chamber $D = C_l$. Set*

$$\hat{r}_{i_k} = \begin{cases} r_{i_k} & \text{if } C_{k-1} \neq C_k, \\ 1 & \text{if } C_{k-1} = C_k. \end{cases}$$

Then

$$D = \hat{r}_{i_1} \cdots \hat{r}_{i_l} C.$$

For all $k = 1, \ldots, l$, the element of W corresponding to the chamber C_k is $\hat{r}_{i_1} \cdots \hat{r}_{i_k}$. In particular, if the gallery Γ does not stutter, then we have $\hat{r}_{i_k} = r_{i_k}$ for all k and Γ is a canonical gallery for the word

$$w = r_{i_1} \cdots r_{i_l}.$$

Proof. The proof is obvious.   □

12.2 Length of Elements in W

If $r_1, \ldots, r_m$ are simple reflections in the finite reflection group W, the *length* of $w \in W$ is defined as the length l of the shortest expression of w in terms of r_i:

$$w = r_{i_1} \cdots r_{i_l}.$$

The one-to-one correspondence between the elements of the reflection group W and chambers of its Coxeter complex $\mathcal{C}$ as described in Theorem 12.2 allows us to give a geometric interpretation of the length function on W.

Theorem 12.4. *If C is the fundamental chamber in $\mathcal{C}$ then*

$$l(w) = \mathrm{gd}(C, wC).$$

12.3 Opposite Chamber

For every chamber D in $\mathcal{C}$ we have a unique *opposite* chamber

$$-D = \{ -\gamma \mid \gamma \in D \}.$$

It is a chamber indeed, since it can be described as the intersection of those half-spaces in Σ that do not contain dD. Alternatively, $-D$ and its panels can be described as follows. Let $P_1, \ldots, P_m$ be the panels of D. If $H_1, \ldots, H_m$ are walls containing the respective panels $P_1, \ldots, P_m$, then $-D$ is the intersection of those half-spaces bounded by H_i that do not contain D. Since the length of a geodesic gallery equals the number of walls it intersects (Proposition 3.4), it is easy to see that $-D$ can be characterized as the chamber in $\mathcal{C}$ that is farthest away from D.

For many applications, it is useful to reformulate this observation in terms of group elements and their lengths.

Since W acts on $\mathcal{C}$ simply transitively (Theorem 11.10), there is a unique element that sends D to $-D$. Assume now that C is the fundamental chamber. If $w \in W$, then the distance from C to wC is the length of w (Theorem 12.4). Hence the element w_0 that sends C to $-C$ is the unique element of greatest length in W. It is natural to call it the *longest element* in W. We summarize our observations:

Theorem 12.5. *There is a unique element $w_0 \in W$ that satisfies the following two equivalent conditions:*

- *The element w_0 has the maximal length in W.*
- *If C is the fundamental chamber then $w_0 C = -C$.*

12.4 Isotropy Groups

We remain in the standard setting of our study: Φ is a root system in $\mathbb{R}^n$, Σ is the corresponding mirror system, and W is the reflection group.

If α is a vector in $\mathbb{R}^n$, its *isotropy group*, or *stabilizer*, or *centralizer* (all these terms are used in the literature) $C_W(\alpha)$ is the group

$$C_W(\alpha) = \{\, w \in W \mid w\alpha = \alpha \,\};$$

if $X \subseteq \mathbb{R}^n$ is a set of vectors, then its *isotropy group* or *pointwise centralizer* in W is the group

$$C_W(X) = \{\, w \in W \mid w\alpha = \alpha \text{ for all } \alpha \in X \,\}.$$

Theorem 12.6. *In this notation,*

(1) *The isotropy group $C_W(X)$ of a set $X \subset \mathbb{R}^n$ is generated by those reflections in W that it contains. In other words, $C_W(X)$ is generated by reflections s_H, $H \in \Sigma$, whose mirrors contain the set X.*

(2) *If X belongs to the closure $\overline{C}$ of the fundamental chamber then the isotropy group $C_W(X)$ is generated by the simple reflections it contains.*

Proof. Consider first the case in which $X = \{\alpha\}$ consists of one vector. Write $W' = C_W(\alpha)$. If the vector α does not belong to any mirror in Σ then it lies in one of the open chambers in $\mathcal{C}$, say D, and $wD = D$ for any $w \in W'$. It follows from the simple transitivity of W on the Coxeter complex $\mathcal{C}$ (Theorem 11.6) that $W' = 1$, and the theorem is true since W' contains no reflections.

Now denote by Σ' the set of all mirrors in Σ that contain α. Obviously, Σ' is a closed mirror system and is invariant under the action of W'.

Consider the set $\mathcal{C}'$ of all chambers D such that $\alpha \in \overline{D}$. Notice that $\mathcal{C}'$ is invariant under the action of W'.

If $D \in \mathcal{C}'$ and P is a panel of D containing α, then the wall H of P belongs to Σ', and the chamber D' adjacent to D via the panel P belongs to $\mathcal{C}'$. Also, if two chambers $D, D' \in \mathcal{C}'$ are adjacent in $\mathcal{C}$ and have the panel P in common, then $\overline{P} = \overline{D} \cap \overline{D'}$ contains α, and the wall H containing the panel P and its closure $\overline{P}$ belongs to Σ'.

These observations allow us to prove that any two chambers D and D' in $\mathcal{C}'$ can be connected by a gallery that belongs to $\mathcal{C}'$. Indeed, let

$$D = D_0, D_1, \ldots, D_l = D'$$

be a geodesic gallery connecting D and D'. If one of the chambers in the gallery, say D_k, does not belong to $\mathcal{C}'$, then select the minimal k with this property and look at the wall H separating D_{k-1} and D_k. The chambers D, D_{k-1}, D' lie on the same side of the wall H as the point α. But a geodesic gallery intersects each wall only once. Hence the entire gallery belongs to $\mathcal{C}'$.

Now take an arbitrary $w \in W'$ and consider a gallery $D_0, \ldots, D_l$ in $\mathcal{C}'$ connecting the chambers $D = D_0$ and $D_l = wD$. If s_i is the reflection in the common panel of the consecutive chambers D_{i-1} and D_i, $i = 1, \ldots, l$, then $D' = s_l \cdots s_1 D$. Since W acts on $\mathcal{C}$ simply transitively, $w = s_l \cdots s_1$. But, for each i, $s_i \in W'$, the group W' is therefore generated by the reflections it contains. This proves (1) in our special case. The statement (2) for $X = \{\alpha\}$ follows from the observation that if $D = C$ is the fundamental chamber, then the proof of Theorem 11.1 can be repeated word for word for W' and $\mathcal{C}'$, showing that W' is generated by reflections in the walls of the fundamental chamber C, i.e., by simple reflections.

Now consider the general case. If every point in X belongs to every mirror in Σ then $C_W(X) = W$ and the theorem is trivially true. Otherwise, take any α in X such that the system Σ' of mirrors containing α is strictly smaller than Σ. Then $C_W(X) \leqslant C_W(\alpha)$, and $W' = C_W(\alpha)$ is itself the reflection group of Σ'. We can use induction on the number of mirrors in Σ, and application of the inductive assumption to Σ' completes the proof.   □

12.5 Parabolic Subgroups

Let Π be a simple system in the root system Φ and $r_1, \ldots, r_m$ the corresponding system of simple reflections. Set $I = \{1, 2, \ldots, m, \}$. For a subset $J \subseteq I$ define

$$W_J = \langle r_i \mid i \in J \rangle;$$

subgroups W_J are called *standard parabolic subgroups* of W. Notice that $W_I = W$ and $W_\emptyset = 1$.

For each $i = 1, \ldots, m$, denote by $\overline{P}_i$ the (closed) panel of the closed fundamental chamber $\overline{C}$ corresponding to the reflection r_i, and set

$$\overline{P}_J = \bigcap_{i \in J} P_i.$$

By virtue of Theorem 12.6,
$$W_J = C_W(\overline{P}_J).$$

We are now in a position to obtain a very easy proof of the following beautiful properties of standard parabolic subgroups.

Theorem 12.7. *If J and K are subsets of I then*

$$W_{J \cup K} = \langle W_J, W_K \rangle$$

and

$$W_{J \cap K} = W_J \cap W_K.$$

Proof. The first equality is obvious, while the second one follows from the observation that
$$W_J \cap W_K = C_W(\overline{P}_J) \cap C_W(\overline{P}_K) = C_W(\overline{P}_J \cup \overline{P}_K).$$

By Theorem 12.6, the latter group is generated by those simple reflections whose mirrors contain the both sets $\overline{P}_J$ and $\overline{P}_K$, that is, by reflections r_i with $i \in J \cap K$. Therefore

$$W_{J \cap K} = W_J \cap W_K.$$

$\square$

As an obvious corollary, Theorem 12.7 gives a list of simple reflections containing in the given standard parabolic subgroup W_J:

$$\{ r_1, \ldots, r_n \} \cap W_J = \{ r_i \mid i \in J \}.$$

We have an important geometric interpretation of this result.

Proposition 12.8. *Let D and E be the chambers corresponding to the elements u and v of a standard parabolic subgroup P_J. If D and E are r_j-adjacent then $j \in J$.*

Proof. Since D and E are r_j-adjacent, then by Theorem 12.2, we have $ur_j = v$ and $r_j \in P_J$. Therefore $j \in J$. $\square$

Exercises

LENGTH OF ELEMENTS.

12.1. Prove that the length of the permutation

$$w = \begin{pmatrix} 1 & 2 & \cdots & n-1 & n \\ i_1 & i_2 & \cdots & i_{n-1} & i_n \end{pmatrix}$$

with respect to the system of standard generators

$$(12), (23), \ldots, (n-1, n)$$

of Sym_n is the number of inversions in w, that is, the number of pairs (j, k) such that $j < k$ and $i_j > i_k$.

12.2. Prove that the longest element in Sym_n with respect to the system of standard generators

$$(12), (23), \ldots, (n-1, n)$$

is the permutation

$$\begin{pmatrix} 1 & 2 & \cdots & n-1 & n \\ n & n-1 & \cdots & 2 & 1 \end{pmatrix}.$$

12.3. Observe that the longest element $w_0 \in W$ is an involution.

12.4. Find the longest elements in the groups $G_2(n)$, BC_n, D_n.

12.5. Let I be the identity transformation of $\mathbb{R}^n$ and $-I$ the transformation that sends every vector α to $-\alpha$. Prove that the transformation

$$-w_0 = -I \cdot w_0$$

sends simple vectors to simple vectors and therefore acts as a permutation (possibly the identity permutation) of the system Π of simple roots. Find this permutation for reflection groups $W = A_n, BC_n, D_n$.

12.6. Show that the length of the longest element in the finite reflection group W equals the number of mirrors in its mirror system Σ.

12.7. Let Φ^+ be the system of positive roots and $R = \{\, r_1, \ldots, r_m \,\}$ the corresponding system of simple reflections in W. Prove that the length of $w \in W$ with respect to R can be expressed as the number of positive roots sent by w to negative roots:

$$l(w) = \left| w\Phi^+ \cap \Phi^- \right|.$$

ISOTROPY GROUPS.

12.8. For the symmetry group of the cube $\Delta = [-1, 1]^3$, find the isotropy groups
(a) of a vertex of the cube,
(b) of the midpoint of an edge,
(c) of the center of a 2-dimensional face.

12.9. Let Φ be the root system of the finite reflection group W and $\alpha \in \Phi$. Prove that the isotropy group $C_W(\alpha)$ is generated by the reflections s_β for all roots $\beta \in \Phi$ orthogonal to α.

12.10. The *centralizer* $C_W(u)$ of an element $u \in W$ is the set of all elements in W that commute with u:

$$C_W(u) = \{\, v \in W \mid vu = uv \,\}.$$

Let s_α be the reflection corresponding to the root $\alpha \in \Phi$. Prove that

$$C_W(s_\alpha) = \langle s_\alpha \rangle \times \langle\, s_\beta \mid \beta \in \Phi \text{ and } \beta \text{ orthogonal to } \alpha \,\rangle.$$

12.11. Let $W = \mathrm{Sym}_n$ and $r = (12)$. Prove that

$$C_W(r) = \langle (12) \rangle \times \langle (34), (45), \ldots, (n-1, n) \rangle$$

and that $C_W(r)$ is isomorphic to $\mathrm{Sym}_2 \times \mathrm{Sym}_{n-2}$.

12.12. Let Δ be a convex polytope and assume that its group of symmetries contains a subgroup W generated by reflections. If Γ is a face of Δ, prove that the setwise stabilizer of Γ in W,

$$\mathrm{Stab}_W(\Gamma) = \{\, w \in W \mid w\Gamma = \Gamma \,\},$$

is generated by reflections.

12.13. INVOLUTIONS IN REFLECTION GROUPS. Prove that if t is an involution in a finite reflection group W then t is a product of pairwise commuting reflections.

12.14. Let $W = A_{n-1}$ and let us view W as the symmetric group Sym_n of the set $[n]$, so that the simple reflections in W are

$$r_1 = (12), r_2 = (23), \ldots, r_{n-1} = (n-1, n).$$

Prove that the standard parabolic subgroup

$$P = \langle r_1, \ldots, r_{k-1}, r_{k+1}, \ldots, r_{n-1} \rangle$$

is the stabilizer in Sym_n of the set $\{\, 1, \ldots, k \,\}$ and thus is isomorphic to $\mathrm{Sym}_k \times \mathrm{Sym}_{n-k}$.

13

Residues

Residues are parabolic subgroups in geometric disguise. In this chapter, the reader will see that the systematic use of geometric language for the description of properties of parabolic subgroups and their cosets is both natural and efficient.

13.1 Residues

Let $W_J = \langle\, r_i \mid i \in J \,\rangle$ be a standard parabolic subgroup in W. By Proposition 12.8, if C and D are the chambers corresponding to two elements of a parabolic subgroup W_J and C and D are r_j-adjacent, then $j \in J$.

We now introduce on C an equivalence relation $\sim_J$ by setting $C \sim_J D$ if C and D can be connected by a gallery $C = C_0, C_1, \ldots, C_l = D$ such that consecutive chambers C_i and C_{i+1} are r_j-adjacent for some $j \in J$. Then Proposition 12.8 immediately yields that the set C_J of chambers corresponding to elements in W_J constitutes an equivalence class.

We shall call equivalence classes of $\sim_J$ J-*residues*, or *residues* if we do not wish to specify the set of indices J. The residue C_J of the fundamental chamber will be called the *standard J-residue*. Since the action of W on C preserves adjacency, the sets wC_J for arbitrary $w \in W$ are also J-residues; since they cover C, every J-residue has form wC_J for some $w \in W$. Of course, the residue wC_J is the set of chambers corresponding to elements in the left coset wW_J of W_J. Hence its setwise stabilizer in W is the parabolic subgroup wW_Jw^{-1}. (Recall that a *parabolic subgroup* is a conjugate of a standard parabolic subgroup.)

Let F be a face of Σ, the hyperplane arrangement of C. We will say the F is a *J-face* if

$$F = \bigcap_{k=1}^{l} H_{i_k},$$

where H_{i_k} is a hyperplane in Σ labeled i_k, and $J = \{\, i_1, i_2, \ldots, i_l \,\}$.

We shall identify the residue wC_J and the coset wW_J.

A.V. Borovik and A. Borovik, *Mirrors and Reflections: The Geometry of Finite Reflection Groups*, Universitext, DOI 10.1007/978-0-387-79066-4_13,

Hence we have natural one-to-one correspondences between the four classes of objects, for J a subset of I:

- J-faces of Σ;
- parabolic subgroups conjugate to the standard parabolic subgroup W_J;
- J-residues;
- left cosets of W with respect to W_J.

13.2 Example

All of the nontrivial residues of C_3 are shown in Figure 13.1. The face of a residue is the unique largest face of Σ that is common to all the chambers that constitute the residue.

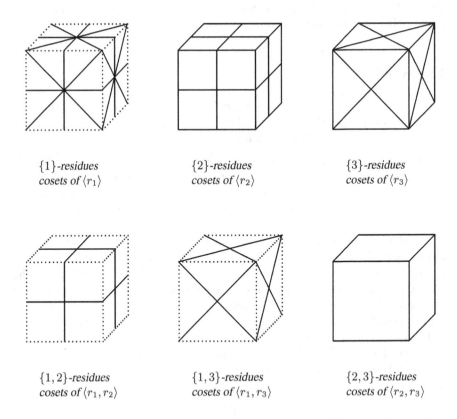

$\{1\}$-residues
cosets of $\langle r_1 \rangle$

$\{2\}$-residues
cosets of $\langle r_2 \rangle$

$\{3\}$-residues
cosets of $\langle r_3 \rangle$

$\{1,2\}$-residues
cosets of $\langle r_1, r_2 \rangle$

$\{1,3\}$-residues
cosets of $\langle r_1, r_3 \rangle$

$\{2,3\}$-residues
cosets of $\langle r_2, r_3 \rangle$

Fig. 13.1. The residues of C_3. The residues in each case are separated by heavy lines, and are to be interpreted as equivalence classes of chambers from Figure 12.1.

13.3 The Mirror System of a Residue

Let $\mathcal{A}$ be a residue with the setwise stabilizer $W_\mathcal{A}$. We call a panel P *internal* in $\mathcal{A}$ if P is a common panel of two distinct adjacent chambers C and D in $\mathcal{A}$. A wall H containing an internal panel P is also called *internal*. If s is a reflection in H, then $D = sC$, and since $\mathcal{A}$ is a left coset of a parabolic subgroup, $s\mathcal{A} = \mathcal{A}$.

Lemma 13.1. *If chambers C and D of a residue $\mathcal{A}$ lie on opposite sides of a wall H, then H is an internal wall of $\mathcal{A}$.*

Proof. The chambers C and D can be connected by a gallery

$$C = C_0, \ldots, C_l = D$$

that lies in $\mathcal{A}$. The wall H intersects the gallery in the sense made explicit in Lemma 3.5: H contains a common panel of two consecutive distinct chambers C_i and C_{i+1}, hence is internal in $\mathcal{A}$.   □

Lemma 13.2. *If s is a reflection in a wall H such that $s\mathcal{A} = \mathcal{A}$, then H is an internal wall of $\mathcal{A}$.*

Proof. Indeed, $\mathcal{A}$ cannot lie on one side of H; hence the result follows from the previous lemma.   □

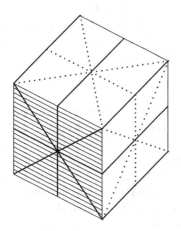 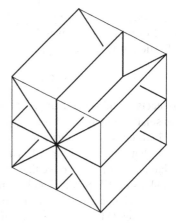

Fig. 13.2. The mirror system of a residue (Theorem 13.3). Shaded is a $\{2, 3\}$-residue (a coset of $\langle r_2, r_3 \rangle$) in the notation of Figure 13.1.

The following statement is now obvious (Figure 13.2).

Theorem 13.3. *Internal walls of a residue $\mathcal{A}$ form a mirror system $\Sigma_\mathcal{A}$. Its reflection group is the parabolic subgroup $W_\mathcal{A}$.*

13.4 Residues are Convex

Theorem 13.4. *Residues are convex.*

Proof. Let C and D be two chambers in a residue $\mathcal{A}$ and $C = C_0, \ldots, C_l = D$ a geodesic gallery that connects C and D. Then every panel between consecutive chambers C_i and C_{i+1} in the gallery belongs to a wall that separates C and D (Corollary 3.6), that is, to an internal wall. Now we can prove by induction, starting from the chamber C_0, that every chamber C_{i+1} is obtained from a chamber $C_i \in \mathcal{A}$ by reflecting in an internal wall of $\mathcal{A}$ and hence also belongs to $\mathcal{A}$. $\square$

13.5 Residues: the Gate Property

Theorem 13.5. *Let C be a chamber and $\mathcal{A}$ a residue in C. Then $\mathcal{A}$ contains a unique chamber G (called a* gate *or C-gate) such that for any other chamber $D \in \mathcal{A}$, there is a geodesic gallery connecting C and D and passing through G.*

Proof. If C belongs to $\mathcal{A}$, then C is obviously a C-gate in $\mathcal{A}$. Therefore we can assume that C does not belong to $\mathcal{A}$.

Let G be a chamber in $\mathcal{A}$ with the shortest distance to C. First we want to prove that G is uniquely determined by this requirement. For that purpose we wish to check first that C and G lie on the same side with respect to any internal wall of $\mathcal{A}$. Assuming the contrary, let H be an internal wall that separates C and G. Then by Lemma 3.5, H intersects a geodesic gallery $C = C_0, \ldots, C_l = G$ in the sense that the common panel of two consecutive chambers C_i and C_{i+1} belongs to H. If $s = s_H$ is the reflection in H, then $sC_{i+1} = C_i$ and $sG \in \mathcal{A}$. Hence the gallery

$$C = C_0, C_1, \ldots, C_i = sC_{i+1}, sC_{i+2}, \ldots, sC_l = sG$$

connects C and sG and, after deletion of one of the repeated chambers C_i and sC_{i+1}, has smaller length. Therefore $\mathrm{gd}(C, G) > \mathrm{gd}(C, sG)$, contrary to our choice of G.

Now consider the system Σ^* of internal walls of $\mathcal{A}$. It is a subsystem of Σ in the sense that every mirror in Σ^* is a mirror in Σ. Every chamber B of Σ is a subset of a chamber B of Σ^*. Notice that distinct chambers of the residue $\mathcal{A}$ belong to distinct chambers of Σ^*. Since C and G are not separated by internal walls of $\mathcal{A}$, they lie in the same chamber of Σ^*. Hence the chamber G is uniquely determined.

Now let D be an arbitrary chamber in $\mathcal{A}$. Consider the geodesic gallery

$$C = C_0, C_1, \ldots, C_k = G$$

connecting C and G and a geodesic gallery

$$G = C_{k+1}, \ldots, C_l = D$$

connecting G and D. We want to prove that together they form a geodesic gallery connecting C and D. To show this, it is enough to prove that every wall H that separates C and D intersects the gallery $C_0, \ldots, C_k$ only once. If H is internal in $\mathcal{A}$, it cannot separate C and G. Since the gallery $C = C_0, \ldots, C_k = G$ is geodesic, we conclude that H does not intersect it. Hence H intersects, and only once, the geodesic gallery $G = C_k, \ldots, C_l = D$. If H is not an internal wall of $\mathcal{A}$, then H cannot intersect the geodesic gallery $G = C_k, \ldots, C_l = D$. Therefore it intersects, at most once, the geodesic gallery $C = C_0, \ldots, c_k = D$. Now by Proposition 3.6, the gallery $C_0, \ldots, C_k, \ldots, C_l$ is geodesic.   □

As a corollary of the proof of Theorem 13.5, we have the following result.

Theorem 13.6. *If $\mathcal{A}$ is a residue in C, $\Sigma_{\mathcal{A}}$ its mirror system, and C^* the set of chambers of the hyperplane arrangement $\Sigma_{\mathcal{A}}$, then every chamber C^* in C^* contains a unique chamber C of $\mathcal{A}$, and conversely, every chamber $C \in \mathcal{A}$ is contained in a chamber C^* in C^**

13.6 The Opposite Chamber

Theorem 13.6 allows us, slightly abusing language, to treat a residue $\mathcal{A}$ as a Coxeter complex for its reflection group $W_{\mathcal{A}}$. In particular, for every chamber D in $\mathcal{A}$ we have a unique *opposite* chamber $-D$; it can be defined as follows. Let $P_1, \ldots, P_m$ be those panels of D that are internal in A. If $H_1, \ldots, H_m$ are walls containing the respective panels $P_1, \ldots, P_m$, then the intersection of those half spaces bounded by H_i that do not contain d is a chamber in $\Sigma_{\mathcal{A}}$. By Theorem 13.6 this intersection contains a chamber $-D \in \mathcal{A}$. Since $\mathcal{A}$ is convex (Theorem 13.4), a geodesic gallery connecting D and $-D$ belongs to $\mathcal{A}$ and intersects every internal wall in $\mathcal{A}$. Since the length of a geodesic gallery equals the number of walls it intersects (Proposition 3.4), it is easy to see that $-D$ can be characterized as the chamber in A that is farthest away from D.

Theorem 13.7. *Let C be a chamber and $\mathcal{A}$ a residue in C. Let G be a C-gate in $\mathcal{A}$ and $-G$ the chamber in $\mathcal{A}$ opposite to G.*

(a) *If D is an arbitrary chamber in $\mathcal{A}$, then there is a geodesic gallery that connects C and $-G$ and passes through G and D.*

(b) *The chamber $-G$ can be characterized by the following property: it is the only chamber in $\mathcal{A}$ such that all chambers D in $\mathcal{A}$ adjacent to it have smaller distance to C: $\mathrm{gd}(C, D) < \mathrm{gd}(C, -G)$.*

Proof. (b) is an immediate corollary of (a). To prove (a), one needs only to observe that the concatenation of a geodesic gallery from C to D via G (which exists by the gate property, Theorem 13.5) and a geodesic gallery from D to $-G$ is the required geodesic gallery.   □

Exercises

The aim of these exercises is to translate Theorem 13.5 into the language of parabolic subgroups and the length function.

Let W be a finite reflection group generated by a simple system of reflections $R = \{r_1, \ldots, r_m\}$ and P a standard parabolic subgroup. Let $w \in W$.

13.1. Prove that the coset wP contains a unique element w' of minimal length.

13.2. Furthermore, every element $v \in wP$ can be written as $v = w'v'$, where $v' \in P$ and

$$l(w) = l(w') + l(v').$$

13.3. Prove that the coset wP contains a unique element w'' of maximal length.

13.4. Show that $(w')^{-1}w''$ is the longest element in P (with respect to the system of simple reflections $P \cap R$).

13.5. Let $W = A_{n-1}$; we view W as the symmetric group Sym_n of the set $[n]$, so that the simple reflections in W are

$$r_1 = (12), r_2 = (23), \ldots, r_{n-1} = (n-1, n).$$

Consider the parabolic subgroup

$$P = \langle r_1, \ldots, r_{k-1}, r_{k+1}, \ldots, r_{n-1} \rangle.$$

We know (Exercise 12.14) that P is the stabilizer in Sym_n of the set $\{1, \ldots, k\}$. Let $w \in \mathrm{Sym}_n$ be an arbitrary permutation. Prove that the element of minimal length in the coset wP is

$$w' = \begin{pmatrix} 1 & 2 & \cdots & n-1 & n \\ i_1 & i_2 & \cdots & i_{n-1} & i_n \end{pmatrix},$$

where $i_1 < i_2 < \cdots < i_k$ are the numbers $w(1), \ldots, w(k)$ written in increasing order and $i_{k+1} < \cdots < i_n$ are the numbers $w(k+1), \ldots, w(n)$ also written in increasing order.

Generalized Permutahedra

Let Σ be a closed mirror system in the real Euclidean space V. We say that a point $\alpha \in V$ is in *general position* if α does not belong to any mirror from Σ.

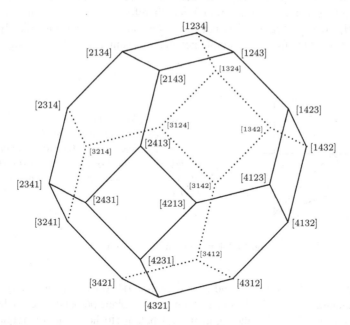

Fig. 14.1. A permutahedron for the group $A_3 = \mathrm{Sym}_4$. Its vertices form one orbit under the permutation action of Sym_4 in $\mathbb{R}^3$ and can be labeled by elements of Sym_4. Here $[i_1 i_2 i_3 i_4]$ denotes the permutation $1 \mapsto i_1, \ldots, 4 \mapsto i_4$.

Let δ be any point in general position, $W \cdot \delta$ its orbit under the reflection group W associated with Σ, and Δ the convex hull of $W \cdot \delta$. We shall call Δ a *generalized permutahedron* (Figure 14.1) and study it in some detail.

A.V. Borovik and A. Borovik, *Mirrors and Reflections: The Geometry of Finite Reflection Groups*, Universitext, DOI 10.1007/978-0-387-79066-4_14,
© Springer Science+Business Media, LLC 2010

Theorem 14.1. *In the notation above, the following statements hold.*

(1) *Vertices of Δ are exactly all points in the orbit $W \cdot \delta$, and each chamber in C contains exactly one vertex of Δ.*
(2) *Every edge of Δ is parallel to some vector in Φ and intersects exactly one wall of the Coxeter complex C.*
(3) *The edges emanating from the given vertex are directed along roots forming a simple system.*
(4) *If α is the vertex of Δ contained in a chamber C then the vertices adjacent to α are exactly all the mirror images $s_i \alpha$ of α in walls of C.*

Proof. Notice, first of all, that since all points in the orbit $W \cdot \delta$ lie at the same distance from the origin, they belong to some sphere centered at the origin. Therefore points in $W \cdot \delta$ are the vertices of the convex hull of $W \cdot \delta$. Next, because of simple transitivity of W on the Coxeter complex C, every chamber in C contains exactly one vertex of Δ. This proves (1).

Let now α and β be two adjacent, i.e., connected by an edge, vertices of Δ. Then β belongs to a chamber distinct from α, and therefore the edge $[\alpha, \beta]$ intersects some mirror H_ρ. If the edge $[\alpha, \beta]$ is not perpendicular to H_ρ, we immediately have a contradiction to the following simple geometric argument (see Figure 14.2).

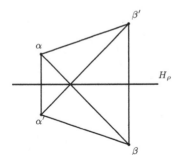

The segment $[\alpha, \beta]$ not normal to a mirror H_ρ that it crosses cannot be an edge of the permutahedron Δ; indeed, if α' and β' are reflections of α and β in H_ρ then α' and β' are also vertices of Δ, and $[\alpha, \beta]$ belongs to the convex hull of $\alpha, \beta, \alpha', \beta'$.

Fig. 14.2. For the proof of Theorem 14.1.

In Figure 14.2, the points α' and β' are symmetric to α, β, respectively, and the convex quadrangle $\alpha\alpha'\beta\beta'$ lies in a 2-dimensional plane perpendicular to the mirror of symmetry H_ρ. Therefore the segment $[\alpha, \beta]$ belongs to the interior of the quadrangle and cannot be an edge of Δ.

Hence $[\alpha, \beta]$ is perpendicular to H_ρ, and hence $\beta - \alpha = c\rho$ for some c and the mirror H_ρ is uniquely determined, which proves (2).

Now select a linear function f that attains its minimum on Δ at the point α and does not vanish at roots in Φ. Let Φ^+ and Π be the positive and simple systems in Φ associated with f. If $s_{\pm\rho} = s_\rho = s_{-\rho}$ is the reflection in W for the roots $\pm\rho$, then $s_{\pm\rho}\alpha$ is a vertex of Δ and $f(s_{\pm\rho}\alpha - \alpha) > 0$. But $s_{\pm\rho}\alpha - \alpha = c\rho$ for some

c. After swapping notation for $+\rho$ and $-\rho$ we can assume without loss of generality that $f(\rho) > 0$, i.e., $\rho \in \Phi^+$ and $c > 0$. Let $\beta_1, \dots, \beta_m$ be all vertices of Δ adjacent to α. Then $\beta_i - \alpha = c_i \rho_i$ for some $\rho_i \in \Phi^+$ and $c_i > 0$.

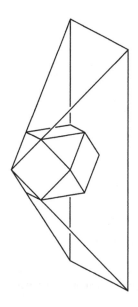

One of the simple principles of linear programming that is extremely useful in the study of Coxeter groups: a convex polytope is contained in the convex polyhedral cone spanned by the edges emanating from the given vertex.

Fig. 14.3. For the proof of Theorem 14.1.

And here comes the punch line: notice that the convex polytope Δ is contained in the convex cone Γ spanned by the edges emanating from α (Figure 14.3). Since every positive root $\rho \in \Phi^+$ points from the vertex α to the vertex $s_\rho \alpha$ of Δ, all positive roots lie in the convex cone spanned by the roots $\rho_i \in \Phi^+$ pointing from α to vertices β_i adjacent to α. But this means exactly that the ρ_i form the simple system Π in Φ^+, which proves (3). Also, the fact that $\beta_i - \alpha = c\rho_i$ for $c > 0$ means that $\alpha \in V_{\rho_i}^-$. Since this holds for all simple roots, α belongs to the fundamental chamber

$$C = \bigcap V_{\rho_i}^-$$

(Theorem 10.1). But by the same theorem, C is bounded by the mirrors of simple reflections and $\beta_i = s_{\rho_i}\alpha$ is the mirror image of α in the wall H_{ρ_i} containing a panel of C. Thus (4) is also proven. $\qquad\square$

Exercises

14.1. Sketch permutahedra for the reflection groups

$$A_1 \oplus A_1, \; A_2, \; BC_2, \; A_1 \oplus A_1 \oplus A_1, \; A_2 \oplus A_1, \; BC_2 \oplus A_1.$$

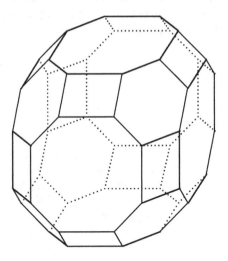

Fig. 14.4. A permutahedron for BC_3 (Exercise 14.2).

14.2. Label, in a way analogous to Figure 14.1, the vertices of a permutahedron for the hyper-octahedral group BC_3 (Figure 14.4) by elements of the group.

14.3. Let Δ be a permutahedron for a reflection group W. Prove that there is a one-to-one correspondence between faces of Δ and residues in the Coxeter complex $\mathcal{C}$ of W. Namely, the set of chambers containing vertices of a given face is a residue.

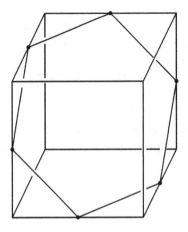

Fig. 14.5. Regular hexagon as a cross section of the cube.

14.4. As you can see in Figure 14.5, there is a cross section of the cube that has the shape of a regular hexagon. Show that there is a cross section of the 4-dimensional cube $[0, 1]^4$ by a 3-dimensional hyperplane that has the shape of a permutahedron for the reflection group A_3.

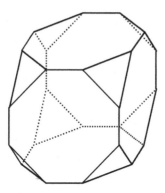

Fig. 14.6. Truncated cube.

14.5. Describe the truncated cube (Figure 14.6) as the closure of an orbit under the action of the appropriate reflection group.

Part IV

Classification

15

Generators and Relations

15.1 Reflection Groups are Coxeter Groups

Let W be a finite reflection group and $R = \{r_1, \ldots, r_m\}$ the set of simple reflections in W. Denote by m_{ij} the orders of pairwise products of simple reflections:

$$m_{ij} = |r_i r_j|.$$

Notice that $m_{ii} = 1$ for all i. We shall soon see that the numbers m_{ij} play a crucial role in our theory.

Theorem 15.1. *The group W is given by the following generators and relations:*

$$W = \langle\, r_1, \ldots, r_m \mid (r_i r_j)^{m_{ij}} = 1 \,\rangle.$$

Our proof closely follows ideas from Grove and Benson [GB, Chapter 6].

But first we need to explain the meaning of the terminology used in the statement of the theorem. Notice that the relations $(r_i r_j)^{m_{ij}} = 1$ are obviously satisfied in W. What we claim is that any other relation $r_{i_1} r_{i_2} \cdots r_{i_l} = 1$ is a corollary of the relations $(r_i r_j)^{m_{ij}} = 1$ in the following sense:

Given a word $w = r_{i_1} r_{i_2} \cdots r_{i_l}$ that equals 1 in W, this word can be transformed in the empty word by the successive application, when appropriate, of the following two operations:

(∗)   We delete from w an occurrence of a twice repeated generator $r_i r_i$. In other words, we apply the relation $r_i r_i = 1$.

(∗∗)   We replace a subword $r_i r_j$ by $(r_j r_i)^{m_{ij}-1}$. In other words, we apply the relation

$$r_i r_j = (r_j r_i)^{m_{ij}-1},$$

which is, of course, a consequence of $(r_i r_j)^{m_{ij}} = 1$.

A.V. Borovik and A. Borovik, *Mirrors and Reflections: The Geometry of Finite Reflection Groups*, Universitext, DOI 10.1007/978-0-387-79066-4_15,
© Springer Science+Business Media, LLC 2010

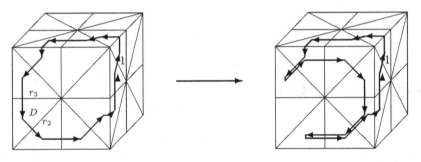

Removing a chamber D from a circular gallery. Here we use the relation $r_3 r_2 = r_2 r_3 r_2 r_3 r_2 r_3$, which is a consequence of $(r_2 r_3)^4 = 1$.

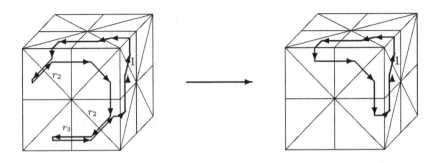

Removing dead end and repeated chambers from a circular gallery. We use the relations $r_2^2 = r_3^2 = 1$.

Fig. 15.1. For the proof of Theorem 15.1.

In general, a *Coxeter group* is a group given by generators and relations

$$W = \langle r_1, \ldots, r_m \mid (r_i r_j)^{m_{ij}} = 1 \rangle$$

with $m_{ii} = 1$ (so that the generators r_i are involutions). It is convenient to allow some values m_{ij} to be ∞, which is a shorthand way of saying that there are no relations between r_i and r_j. Theorem 15.1 says that

every finite reflection group is a Coxeter group.

The converse is also true, although we do not prove it in this book.

Theorem 15.2. *If*

$$W = \langle r_1, \ldots, r_m \mid (r_i r_j)^{m_{ij}} = 1 \rangle$$

is a finite Coxeter group then it is isomorphic to a finite reflection group; moreover, in this isomorphism the generators r_i are represented by reflections.

15.2 Proof of Theorem 15.1

The idea of the proof is best illustrated by Figure 15.1. Let

$$w = r_{i_1} r_{i_2} \cdots r_{i_l} = 1$$

and let

$$\Gamma = (C_0, C_1, \ldots, C_l)$$

be a canonical w-gallery that starts and ends at the fundamental chamber $C = C_0 = C_l$. By applying the relations (∗), we can assume that the gallery Γ has no repeated chambers $C_i = C_{i+1}$. Now let $D = C_m$ be one of the chambers in the gallery that lies at the greatest distance from C. Since

$$\mathrm{gd}(C, C_{m-1}) = \mathrm{gd}(C, D) \pm 1$$

by Proposition 3.8, we obviously have

$$\mathrm{gd}(C, C_{m-1}) = \mathrm{gd}(C, D) - 1,$$

and analogously,

$$\mathrm{gd}(C, C_{m+1}) = \mathrm{gd}(C, D) - 1.$$

Assume that the chambers C_{m-1} and D are r_i-adjacent and the chambers C_{m+1} and D are r_j-adjacent. If we now consider the (r_i, r_j)-residue $\mathcal{R}$ of D, we see that D is farther away from the fundamental chamber C than the chambers C_{m-1} and C_{m+1}, its two neighbors in the residue. Hence, by Theorem 13.7, D is the chamber of the residue $\mathcal{R}$ most distant from the fundamental chamber. If we replace the chambers C_{m-1}, D, C_m in Γ by the sequence

$$C_{m-1},\ C_{m-1}r_j,\ C_{m-1}r_j r_i,\ \ldots,\ C_{m-1}(r_j r_i)^{m_{ij}-1} = C_{m+1},$$

of chambers in $\mathcal{R}$ we get another gallery Γ' with the following properties:

- Γ' starts and ends at C,
- the maximum distance from C to a chamber in Γ' is at most $\mathrm{gd}(C, D)$, and
- the number of chambers in Γ' lying at distance $\mathrm{gd}(C, D)$ from C is strictly smaller than the analogous number for Γ.

Notice that the described transformation is nothing more than an application of (∗∗). Obviously, a successive application of these transformations to our gallery contracts the gallery to a gallery that consists of just one chamber C and corresponds to the empty word in the generators $r_1, \ldots, r_n$. □

Exercises

15.1. Apply the contraction procedure of the proof of Theorem 15.1 to the word

$$r_1 r_2 r_3 r_2 r_3 r_2 r_1 r_2 r_3 r_2 r_3 r_2 r_3 r_1 r_3 r_2 r_3 r_2 r_3 r_2$$

in BC_3.

15.2. Theorem 15.1 can be generalized to *affine reflection groups*, that is, reflection groups of infinite but locally finite mirror systems. Here, a mirror system Σ in $\mathbb{AR}^n$ is called *locally finite* if every ball

$$B(a, r) = \{\, x \in \mathbb{AR}^n \mid d(x, a) < r \,\}$$

intersects only finitely many mirrors in Σ.

1. Check that the proof of Theorem 15.1 can be transferred to reflection groups of Figure 6.2 on page 43.
2. Find generators and relations of these groups.

16

Classification of Finite Reflection Groups

Our treatment of the classification of finite reflection groups closely follows the classical exposition of the theory in Humphreys [Hum, Chapter 2].

16.1 Coxeter Graph

By Theorem 15.1, a finite reflection group W is given by the following generators and relations:
$$W = \langle\, r_1, \ldots, r_m \mid (r_i r_j)^{m_{ij}} = 1 \,\rangle,$$
where $R = \{\, r_1, \ldots, r_m \,\}$ is the set of simple reflections in W and $m_{ij} = |r_i r_j|$. Notice that $m_{ii} = 1$ for all i.

Now we wish to associate with W and a system of simple reflections R a graph G, called a *Coxeter graph*, whose nodes are in one-to-one correspondence with the simple reflections $r_1, \ldots, r_n$ in R. If r_i and r_j are two distinct reflections, then if $m_{ij} = |r_i r_j| > 2$, the nodes r_i and r_j are connected by an edge with mark m_{ij} on it. If $m_{ij} = 2$, that is, if r_i and r_j commute, then there is no edge connecting r_i and r_j.

Since the Coxeter graph G encodes the information about the generators and relations of the finite reflection group W, it determines W as an abstract group.

For example, the graph of the symmetric group $A_{n-1} = \mathrm{Sym}_n$ with respect to the generators
$$(12), (23), \ldots, (n-1, n)$$
is

A.V. Borovik and A. Borovik, *Mirrors and Reflections: The Geometry of Finite Reflection Groups*, Universitext, DOI 10.1007/978-0-387-79066-4_16,
© Springer Science+Business Media, LLC 2010

16.2 Decomposable Reflection Groups

We say that the reflection group W is *indecomposable* or *irreducible* if the graph G is connected; otherwise, W is said to be *decomposable*.

Theorem 16.1. *Assume that W is decomposable and let $G_1, \ldots, G_k$ be the connected components of G. Let R_j be the set of reflections corresponding to nodes in the connected component G_j, $j = 1, \ldots, k$. Let W^j be the parabolic subgroup generated by the set R_j. Then*

$$W = W^1 \times \cdots \times W^k.$$

Proof. If $i \neq j$ then any two reflections $r' \in R_i$ and $r'' \in R_j$ commute; hence

- the subgroups $W^i = \langle R_i \rangle$ and $W^j = \langle R_j \rangle$ commute elementwise.

By Theorem 12.7, the intersection of W^j with the group generated by all W^i with $i \neq j$ is the subgroup generated by $R_j \cap \bigcup_{i \neq j} R_i = \emptyset$, that is, the identity subgroup:

- $W^j \cap \langle W^1, \ldots, \widehat{W^j}, \ldots, W^k \rangle = 1$.

Finally, the subgroups W^i generate W,

- $W = \langle W^1, \ldots, W^k \rangle$.

But these properties of the subgroups W^i mean exactly that

$$W = W^1 \times \cdots \times W^k.$$

$\square$

16.3 Labeled Graphs and Associated Bilinear Forms

We will now classify the finite reflection groups by listing all of their Coxeter graphs.

Define a *labeled graph* to be a finite (undirected) graph whose edges are labeled with integers $\geqslant 3$. If s and t are distinct vertices, let m_{st} denote the label on the edge joining st. It is a convention to omit the label 3 when drawing diagrams: in most important examples, it occurs too frequently. We also make the convention that $m_{st} = 2$ for vertices $s \neq t$ not joined by an edge. Finally, we set $m_{ss} = 1$.

We associate to a labeled graph Γ with vertex set S of cardinality n a symmetric $n \times n$ matrix G by setting

$$g_{st} := -\cos \frac{\pi}{m_{st}}.$$

With every symmetric $n \times n$ matrix $G = G^t$ one can associate a bilinear form $x^t G y \ (x, y \in \mathbb{R}^n)$ and a quadratic form $x^t G x$. The matrix G is called *positive definite* if

$$x^t G x > 0$$

for all $x \neq 0$. We call the graph Γ positive definite when the associated quadratic form is positive definite.

Recall a well-known result from linear algebra: G is positive definite if and only if all its principal minors are positive. Here, the *principal minors* of G are the determinants of the submatrices formed by the first k rows and columns ($0 < k \leqslant n$).

We easily see that when Γ comes from a finite reflection group W, then the associated bilinear form is, in fact, the standard scalar multiplication in Euclidean space; hence the matrix G is *positive definite*. Our strategy for classifying finite reflection groups is to assemble the list of all possible connected positive definite labeled graphs.

16.4 Classification of Positive Definite Graphs

Our study has reached the point where the classification of finite reflection groups becomes a matter of relatively simple matrix computations. Indeed, the Coxeter graph of W is positive definite. Hence the desired classification of finite reflection groups is an immediate consequence of the following result.

Theorem 16.2. *The connected positive definite labeled graphs are exactly those listed in Figure 16.1.*

Proof. A proof of the theorem is based on the following observation about subgraphs of positive definite graphs. We say that a labeled graph Δ is a *subgraph* of Γ if it can be obtained from Γ by any combination of the following two operations:

- deleting some vertices (with adjacent edges), and
- decreasing some edge labels.

We also say in this situation that Γ *contains* Δ.

Lemma 16.3. *A subgraph Γ' of a positive definite graph Γ is positive definite.*

Proof. We enumerate vertices $1, \ldots, n$ of Γ so that $1, \ldots, k$ are vertices of the subgraph Γ'. The edge labels of Γ' satisfy the inequality $m'_{ij} \leqslant m_{ij}$; therefore

$$g'_{ij} = -\cos\frac{\pi}{m'_{ij}} \geqslant -\cos\frac{\pi}{m_{ij}} = g_{ij}.$$

If Γ' is not positive definite, then there exists a nonzero vector $x \in \mathbb{R}^k$ such that $x^t G' x \leqslant 0$. Form the vector

$$y = (|x_1|, |x_2|, \ldots, |x_k|, 0, \ldots, 0)$$

in $\mathbb{R}^n$. Since $y^t G y > 0$, we have

$$0 < \sum_{1 \leqslant i,j \leqslant n} g_{ij} y_i y_j$$

$$= \sum_{1 \leqslant i,j \leqslant k} g_{ij} |x_i||x_j|$$

$$\leqslant \sum_{1\leqslant i,j\leqslant k} g'_{ij}|x_i||x_j|$$

$$= \sum_{1\leqslant i,j\leqslant k} g'_{ij}|x_ix_j|$$

$$\leqslant \sum_{1\leqslant i,j\leqslant k} g'_{ij}x_ix_j \quad (\text{since } g'_{ij} \leqslant 0 \text{ for } i \neq j)$$

$$\leqslant 0,$$

a contradiction.   □

We leave it to the reader as an exercise (Exercise 16.1) to check that all graphs in Figure 16.1 are positive definite, while the graphs in Figure 16.2 are not. It is easier than one might think because the associated matrices consist mostly of zeros. Here are some hints: number the vertices of Γ so that the last vertex, n, is connected only to one vertex, $n-1$. Then the last row of the matrix G contains only two nonzero entries, and $\det G$ can be expanded with respect to the last row, allowing for the inductive argument. To get rid of awkward denominators, it is convenient to carry out the actual computation for the matrix $2G$. The resulting values of $\det 2G$ for the graphs in Figure 16.1 are given in the following table:

A_n	BC_n	D_n	E_6	E_7	E_8	F_4	$G_2(m)$	H_3	H_4
$n+1$	2	4	3	2	1	1	$4\sin^2(\pi/m)$	$3-\sqrt{5}$	$\frac{1}{2}(7-3\sqrt{5})$

In Figure 16.2, the matrices for graphs $\tilde{A}_n$–$\tilde{G}_2$ have zero determinants, while the determinants for $\tilde{H}_3$ and $\tilde{H}_4$ are negative.

The rest of Exercise 16.1 completes the proof of the theorem: a very straightforward argument shows that the only connected labeled graphs Γ that contain no subgraphs listed in Figure 16.2 are those listed in Figure 16.1. For example, since $\tilde{A}_n$ is not a subgraph of Γ, the latter contains no cycle.   □

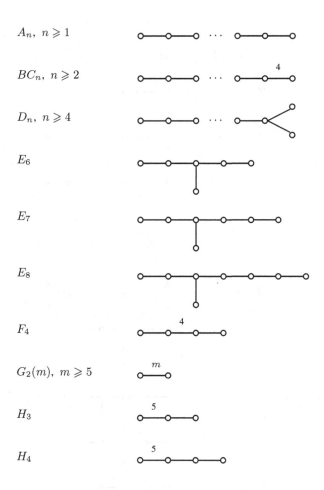

Fig. 16.1. Positive definite graphs.

Exercises

16.1. Check some of the graphs in Figure 16.1, page 121 and Figure 16.2, page 122, for being positive definite or not (resp.). Assuming they all can be checked correctly, complete the proof of Theorem 16.2.

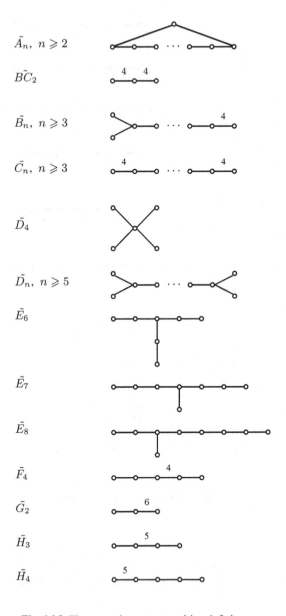

$\tilde{A}_n,\ n \geqslant 2$

$\tilde{BC}_2$

$\tilde{B}_n,\ n \geqslant 3$

$\tilde{C}_n,\ n \geqslant 3$

$\tilde{D}_4$

$\tilde{D}_n,\ n \geqslant 5$

$\tilde{E}_6$

$\tilde{E}_7$

$\tilde{E}_8$

$\tilde{F}_4$

$\tilde{G}_2$

$\tilde{H}_3$

$\tilde{H}_4$

Fig. 16.2. These graphs are not positive definite.

17

Construction of Root Systems

Arguably, this is the dullest chapter of the book. We construct, for each Coxeter graph of type

$$A_n, \ B_n, \ C_n, \ D_n, \ E_6, \ E_7, \ E_8, \ F_4, \ G_2,$$

a root system and briefly list their key properties. As a rule, we do not give any proofs; all the properties mentioned can be checked by direct, although sometimes tedious, calculations. Our exposition follows the classical treatise ny Bourbaki [Bou].

The systems A_n, B_n, C_n, D_n have already been treated in previous chapters. Here, we give a brief summary and fill in the missing details.

The systems E_6, E_7, E_8, F_4, G_2 are exceptionally beautiful; their importance can be fully appreciated in their applications. However, any discussion of these applications is beyond the scope of the present book.

We do not consider the root systems of type H_3 and H_4. The interested reader may wish to consult the books by Grove and Benson [GB] and Humphreys [Hum], which contain a detailed discussion of these root systems. We mention only that the mirror system associated with the root system of type H_3 is the system of mirrors of symmetry of the regular icosahedron (or its dual polytope, the dodecahedron; see Figure 10.4). A rigorous construction of the icosahedron is given in Chapter 20.

17.1 Root System A_n

Let $\epsilon_1, \dots, \epsilon_{n+1}$ be the standard basis in $\mathbb{R}^{n+1}$,

$$\Phi = \{ \epsilon_i - \epsilon_j \mid i, j = 1, \dots, n+1, i \neq j \},$$
$$\Pi = \{ \epsilon_2 - \epsilon_1, \dots, \epsilon_{n+1} - \epsilon_n \}.$$

Then Φ contains $n(n+1)$ vectors, all of which are of equal length. Denote the simple vectors by

$$\rho_1 = \epsilon_2 - \epsilon_1, \ \rho_2 = \epsilon_3 - \epsilon_2, \ \dots, \ \rho_n = \epsilon_{n+1} - \epsilon_n,$$

and take the root

A.V. Borovik and A. Borovik, *Mirrors and Reflections: The Geometry of Finite Reflection Groups*, Universitext, DOI 10.1007/978-0-387-79066-4_17, © Springer Science+Business Media, LLC 2010

$$\rho_0 = \epsilon_{n+1} - \epsilon_1 = \rho_1 + \rho_2 + \cdots + \rho_n.$$

The root ρ_0 is called the *highest* root because it has, of all positive roots, the longest expression in terms of the simple roots. The highest root plays an exceptionally important role in many applications of the theory of root systems, for example, in the representation theory of simple Lie algebras and simple algebraic groups. In the context of our approach to finite reflection groups via systems of mirrors it will act as the marker for synchronizing the mirror system and root system: in every root system that we consider, the highest root ρ_0 is chosen in such way that $-\rho_0$ belongs to the closure $\overline{C}$ of the fundamental chamber C.

In the following diagram the black nodes form the Coxeter graph for A_n; an extra white node demonstrates the relations of the root $-\rho_0$ to the simple roots. We use the following convention: if α and β are two roots, then their nodes are not connected if $\alpha \cdot \beta = 0$ (and the reflections s_α and s_β commute), and the nodes are connected by an edge if

$$\frac{\alpha \cdot \beta}{|\alpha||\beta|} = -\cos\frac{\pi}{m}$$

and $m \geqslant 3$. In fact, m is the order of the product $s_\alpha s_\beta$, and $m \geqslant 3$ if and only if the reflections s_α and s_β do not commute. If $m > 3$ we write the value of m on the edge.

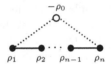

We know that the reflection group W for our root system is the symmetric group Sym_{n+1}, which acts by permuting the vectors ϵ_i.

17.2 Root System $B_n,\ n \geqslant 2$

Let $\epsilon_1, \ldots, \epsilon_n$ be the standard basis in $\mathbb{R}^n$,

$$\Phi = \{\pm\epsilon_i,\ \pm\epsilon_i \pm e_j \mid i, j = 1, \ldots, n, i < j\},$$
$$\Pi = \{\epsilon_1, \epsilon_2 - \epsilon_1, \ldots, \epsilon_n - \epsilon_{n-1}\},$$

Then Φ contains $2n$ short roots $\pm\epsilon_i$ and $2n(n-1)$ long roots $\pm\epsilon_i \pm \epsilon_j,\ i < j$.

It is convenient to enumerate the simple roots as

$$\rho_1 = \epsilon_1,\ \rho_2 = \epsilon_2 - \epsilon_1,\ \ldots,\ \rho_n = \epsilon_n - \epsilon_{n-1}.$$

The highest root is

$$\rho_0 = \epsilon_{n-1} + \epsilon_n.$$

The extended Coxeter diagrams for the system of roots B_2 and B_n with $n \geqslant 3$ are different.

$$\underset{\rho_1}{\bullet}\overset{4}{\rule{1cm}{0.4pt}}\underset{\rho_2}{\bullet}\overset{4}{\cdots\cdots}\underset{-\rho_0}{\circ}$$

Extended Coxeter diagram for B_2.

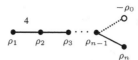

Extended Coxeter diagram for B_n, $n \geqslant 3$.

17.3 Root System C_n, $n \geqslant 2$

Let $\epsilon_1, \ldots, \epsilon_n$ be the standard basis in $\mathbb{R}^n$,

$$\Phi = \{ \pm 2\epsilon_i, \; \pm\epsilon_i \pm e_j \mid i, j = 1, \ldots, n, i < j \},$$
$$\Pi = \{ 2\epsilon_1, \epsilon_2 - \epsilon_1, \ldots, \epsilon_n - \epsilon_{n-1} \}.$$

Then Φ contains $2n$ long roots $\pm 2\epsilon_i$ and $2n(n-1)$ short roots $\pm\epsilon_i \pm \epsilon_j$, $i < j$. We enumerate the simple roots as

$$\rho_1 = 2\epsilon_1, \; \rho_2 = \epsilon_2 - \epsilon_1, \; \ldots, \; \rho_n = \epsilon_n - \epsilon_{n-1}.$$

The highest root is

$$\rho_0 = 2\epsilon_n.$$

$$\underset{-\rho_0}{\circ}\overset{4}{\cdots\cdots}\underset{\rho_1}{\bullet}\overset{4}{\rule{1cm}{0.4pt}}\underset{\rho_2}{\bullet}$$

Extended Coxeter diagram for C_2

$$\overset{4}{\bullet}\!\!-\!\!\bullet\!\!-\!\!\bullet\cdots\bullet\!\!-\!\!\overset{4}{\bullet}\!\cdots\cdots\!\circ$$
$$\rho_1\quad\rho_2\quad\rho_3\quad\rho_{n-1}\quad\rho_n\quad-\rho_0$$

Extended Coxeter diagram for C_n, $n \geqslant 3$.

17.4 Root System D_n, $n \geqslant 4$

Let $\epsilon_1, \ldots, \epsilon_n$ be the standard basis in $\mathbb{R}^n$,

$$\Phi = \{\, \pm\epsilon_i \pm \epsilon_j \mid i, j = 1, 2, \ldots, n, i \neq j \,\};$$

thus D_n is a subsystem of the root system C_n. All roots have the same length. The total number of roots is $2n(n-1)$.

The simple system Π is

$$\rho_1 = \epsilon_1 + \epsilon_2, \ \rho_2 = \epsilon_2 - \epsilon_1, \ \rho_3 = \epsilon_3 - \epsilon_2, \ \ldots, \ \rho_n = \epsilon_n - \epsilon_{n-1}.$$

The highest root is

$$\rho_0 = \epsilon_{n-1} + \epsilon_n.$$

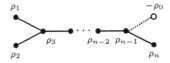

Extended Coxeter diagram for D_n, $n \geqslant 4$.

17.5 Root System E_8

Let $\epsilon_1, \ldots, \epsilon_8$ be the standard basis in $\mathbb{R}^8$,

$$\Phi = \left\{\, \pm\epsilon_i \pm \epsilon_j \quad (i < j), \quad \frac{1}{2}\sum_{i=1}^{8} \pm\epsilon_i \quad (\text{even number of } + \text{ signs}) \,\right\},$$

and for Π take

$$\rho_1 = \frac{1}{2}(\epsilon_1 - \epsilon_2 - \epsilon_3 - \epsilon_4 - \epsilon_5 - \epsilon_6 - \epsilon_7 + \epsilon_8),$$
$$\rho_2 = \epsilon_1 + \epsilon_2,$$
$$\rho_i = \epsilon_{i-1} - \epsilon_{i-2} \quad (3 \leqslant i \leqslant 8).$$

All roots have the same length; the total number of roots is 240.
The highest root is

$$\rho_0 = \epsilon_7 + \epsilon_8.$$

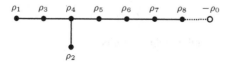

Extended Coxeter diagram for E_8.

17.6 Root System E_7

Take the root system of type E_8 in $\mathbb{R}^8$ just constructed and consider the span V of the roots $\rho_1, \ldots, \rho_7$. Let Φ be the set of 126 roots of E_8 belonging to V:

$$\pm\epsilon_i \pm \epsilon_j \quad (1 \leqslant i < j \leqslant 6), \quad \pm(\epsilon_7 - \epsilon_8), \quad \pm\frac{1}{2}\left(\epsilon_7 - \epsilon_8 + \sum_{i=1}^{6} \pm\epsilon_i\right),$$

where the number of minus signs in the sum is odd.
All roots have the same length. The roots

$$\rho_1, \ldots, \rho_7$$

form a simple system, and the highest root is

$$\rho_0 = \epsilon_8 - \epsilon_7.$$

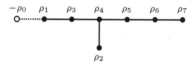

Extended Coxeter diagram for E_7.

17.7 Root System E_6

Again we start with the root system of type E_8 in $\mathbb{R}^8$. Denote by V the span of the roots $\rho_1, \ldots, \rho_6$, and take for Φ the 72 roots of E_8 belonging to V:

$$\pm\epsilon_i \pm \epsilon_j \quad (1 \leqslant i < j \leqslant 5), \qquad \pm\frac{1}{2}\left(\epsilon_8 - \epsilon_7 - \epsilon_6 + \sum_{i=1}^{5} \pm\epsilon_i\right),$$

where the number of minus signs in the sum is odd.

All roots have the same length. The roots

$$\rho_1, \ldots, \rho_6$$

form a simple system, and the highest root is

$$\rho_0 = \frac{1}{2}(\epsilon_1 + \epsilon_2 + \epsilon_3 + \epsilon_4 + \epsilon_5 - \epsilon_6 - \epsilon_7 + \epsilon_8).$$

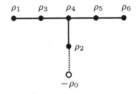

Extended Coxeter diagram for E_6.

17.8 Root System F_4

Let $\epsilon_1, \ldots, \epsilon_4$ be the standard basis in $\mathbb{R}^4$; Φ consists of 24 long roots

$$\pm\epsilon_i \pm \epsilon_j \quad (i < j)$$

and 24 short roots

$$\pm\epsilon_i, \quad \frac{1}{2}(\pm\epsilon_1 \pm \epsilon_2 \pm \epsilon_3 \pm \epsilon_4).$$

For a simple system Π take

$$\rho_1 = \epsilon_2 - \epsilon_3, \ \rho_2 = \epsilon_3 - \epsilon_4, \ \rho_3 = \epsilon_4, \ \rho_4 = \frac{1}{2}(\epsilon_1 - \epsilon_2 - \epsilon_3 - \epsilon_4).$$

The highest root is

$$\rho_0 = \epsilon_1 + \epsilon_2.$$

Extended Coxeter diagram for F_4.

17.9 Root System G_2

Let V be the hyperplane $x_1 + x_2 + x_3 = 0$ in $\mathbb{R}^3$, Φ consists of six short roots

$$\pm(\epsilon_i - \epsilon_j), \quad i < j,$$

and six long roots

$$\pm(2\epsilon_i - \epsilon_j - \epsilon_k),$$

where i, j, k are all different. For a simple system Π take

$$\rho_1 = \epsilon_1 - \epsilon_2, \ \rho_2 = -2\epsilon_1 + \epsilon_2 + \epsilon_3.$$

The highest root is

$$\rho_0 = \epsilon_1 + \epsilon_2.$$

Extended Coxeter diagram for G_2.

17.10 Crystallographic Condition

A root system Φ is called *crystallographic* if for all $\alpha, \beta \in \Phi$,

$$\frac{2\alpha \cdot \beta}{\beta \cdot \beta} \quad \text{is an integer.} \tag{17.1}$$

If you look at the formula for the reflection,

$$s_\alpha \beta = \beta - \frac{2\beta \cdot \alpha}{\alpha \cdot \alpha} \alpha,$$

you can immediately observe that it means that if reflections are represented by matrices in the basis of simple vectors, then all matrix coefficients in these matrices are integers. Since W is generated by reflections, this property holds for all elements in W. Moreover, it is easy to see that the converse is also true:

Proposition 17.1. *Assume that the real vector space V is spanned by a root system Φ for the finite reflection group W. Then Φ is crystallographic if and only if every element of W is represented in the basis of simple roots by a matrix with integer coefficients.*

Theorem 17.2. *The root systems A_n, B_n, C_n, D_n, E_6, E_7, E_8, F_4, G_2 are crystallographic.*

Proof. The proof is a straightforward calculation and is left to the reader as an exercise.

Exercises

For each of the root systems A_n, B_n, C_n, D_n, E_6, E_7, E_8, F_4, G_2:

17.1. Check the crystallographic condition.

17.2. Find the decomposition of the highest root with respect to the simple roots.

17.3. Check, by a direct computation, that the extended Coxeter diagrams are drawn correctly.

17.4. The sets Φ_{long} and Φ_{short} of all long (respectively, short) roots in a root system Φ are root systems on their own. Identify their types when Φ is of type B_n, C_n, or F_4.

CRYSTALLOGRAPHIC ROOT SYSTEMS.

17.5. Prove that in a crystallographic root system, every root can be written as a linear combination of simple roots with integer coefficients.

17.6. For the root systems Φ of types A_2, B_2, C_2, G_2, sketch the sets $\Lambda = \mathbb{Z}\Phi$ of points in $\mathbb{R}^2$ that are linear combinations of roots in Φ with *integer* coefficients,

$$\Lambda = \left\{ \sum_{\alpha \in \Phi} a_\alpha \alpha \mid a_\alpha \in \mathbb{Z} \right\}.$$

Observe that Λ is a *discrete* subgroup of $\mathbb{R}^2$, that is, there is a real number $d > 0$ such that for any $\lambda \in \Lambda$, the circle $\{ \alpha \in \mathbb{R}^2 \mid d(\alpha, \lambda) < d \}$ contains no points from Λ other than λ.

17.7. Prove that if Φ is a crystallographic root system then

$$\Lambda = \left\{ \sum_{\alpha \in \Phi} a_\alpha \alpha \mid a_\alpha \in \mathbb{Z} \right\}$$

is a discrete subgroup of the vector space spanned by Φ.

17.8. Check that root systems of type $G_2(5)$ are not crystallographic. Also, show that root systems of type $G_2(m)$ are not crystallographic when $m \geqslant 7$.

17.9. Prove also that the additive group $\mathbb{Z}\Phi$ generated in $\mathbb{R}^2$ by a root system Φ of type $G_2(5)$ is everywhere dense in $\mathbb{R}^2$.

17.10. GROUPS H_3 AND H_5.

We have not constructed root systems corresponding to the Coxeter diagrams H_3 and H_4. Observe, however, that

1. The mirror systems of mirrors of symmetries of the dodecahedron (equivalently, of the icosahedron) have type H_3 (see Figure 10.4); a construction of the icosahedron will be given in Chapter 20.
2. Root systems H_3 and H_4 contain a subsystem of type $G_2(5)$ and therefore are noncrystallographic (use Exercise 17.8).
3. In any root systems of type H_3 or H_4 all roots are conjugate and therefore have the same length.

Project

The next problem is quite serious; its solution requires a systematic approach; we would recommend it as a long-term research project.

17.11 (Felder and Veselov; Pfeiffer and Roehrle). In finite irreducible reflection groups W, find all involutions t such that $C_W(t)$ is generated by reflections.

18

Orders of Reflection Groups

In this chapter we shall use information about the root systems accumulated in Chapter 17 to determine the orders of the finite reflection groups

$$A_n, \ BC_n, \ D_n, \ E_6, \ E_7, \ E_8, \ F_4, \ G_2.$$

Our exposition follows Humphreys [Hum, Theorem 2.11].

First of all, start by observing that the group A_1 obviously has order 2. The groups A_2, BC_2, G_2 are the dihedral groups of orders 6, 8, 12, respectively.

Let Φ be one of the root systems listed above and W its reflection group. To work out the order of W we need first to study the action of W on Φ.

Lemma 18.1. *The long (respectively, short) roots in Φ are conjugate under the action of W.*

Proof. We know from Theorem 11.11 that every root is conjugate to a simple root. Therefore it will be enough to prove that the simple long (respectively, short) roots are conjugate. Direct observation of Coxeter graphs shows that the nodes for any two simple roots of the same length can be connected by a sequence of edges with marks 3. Hence it will be enough to prove that if ρ_i and ρ_j are distinct simple roots so that $m_{ij} = 3$, then ρ_i and ρ_j are conjugate; but this follows from Exercise 8.7 applied to the planar root system

$$\Phi' = \Phi \cap (\mathbb{Z}\rho_i + \mathbb{Z}\rho_j).$$

$\square$

Theorem 18.2. *The orders of the indecomposable reflection groups are given in the following list:*

$$|A_n| = n!$$
$$|BC_n| = 2^n \cdot n!$$
$$|D_n| = 2^{n-1} \cdot n!$$
$$|E_6| = 2^7 \cdot 3^4 \cdot 5$$

A.V. Borovik and A. Borovik, *Mirrors and Reflections: The Geometry of Finite Reflection Groups*, Universitext, DOI 10.1007/978-0-387-79066-4_18,
© Springer Science+Business Media, LLC 2010

$$|E_7| = 2^{10} \cdot 3^4 \cdot 5 \cdot 7$$
$$|E_8| = 2^{14} \cdot 3^5 \cdot 5^2 \cdot 7$$
$$|F_4| = 2^7 \cdot 3^2$$
$$|G_2| = 12$$

Proof. In all cases the highest root ρ_0 is a long root. Since all long roots are conjugate,

$$|W| = \begin{pmatrix} \text{number} \\ \text{of long} \\ \text{roots} \end{pmatrix} \cdot |C_W(\rho_0)|.$$

On the other hand, $C_W(\rho_0) = C_W(-\rho_0)$. One can easily check, using the formulas of Chapter 17, that $\rho_0 \cdot \rho_i \geqslant 0$ for all simple roots ρ_i; hence $-\rho_0 \cdot \rho_i \leqslant 0$ and the root $-\rho_0$ belongs to the closed fundamental chamber $\overline{C}$. By Theorem 12.6, the isotropy group $C_W(-\rho_0)$ is generated by the simple reflections that fix the root $-\rho_0$. These simple reflections are exactly the reflections for the black nodes on the extended Coxeter graphs in Chapter 17 that are not connected by an edge to the white node $-\rho_0$. Therefore the extended Coxeter graph, with the node $-\rho_0$ and the nodes adjacent to $-\rho_0$ deleted, is the Coxeter graph for $W' = C_W(\rho_0)$, which allows us to determine the isomorphism type and the order of W'.

The rest is a case-by-case analysis.

$\boldsymbol{A_n}$. We know that $|A_1| = 2 = 2!$ and $|A_2| = 6 = 3!$. We want to prove by induction that $|A_n| = (n+1)!$. Φ contains $n(n+1)$ roots (all of them of the same length), and W' is of type A_{n-2}. By the inductive hypothesis,

$$|W'| = [(n-2)+1]! = (n-1)!$$

and

$$|W| = n(n+1) \cdot (n-1)! = (n+1)!.$$

Of course, we know that $W = \mathrm{Sym}_{n+1}$, and there was not much need for a new proof of the fact that $|\mathrm{Sym}_n| = n!$. But we wished to use an opportunity to show how much information about a reflection group is contained in its extended Coxeter graph.

$\boldsymbol{BC_n}$. We know that the root systems B_n and C_n have the same mirror system and reflection group. It will be more convenient for us to compute with the root system C_n. It contains $2n$ long roots, and the Coxeter graph for W' is of type C_{n-1}. Thus

$$|W| = 2n \cdot 2^{n-1}(n-1)! = 2^n n!.$$

$\boldsymbol{D_n}$. All roots are long, and their number is $2n(n-1)$. The group W' has disconnected Coxeter graph with connected components of types D_{n-2} and A_1; hence $W' = W'' \times W'''$, where W'' is of type D_{n-2}, has, by the inductive hypothesis, order $2^{n-3}(n-2)!$, and $|W'''| = 2$. Therefore

$$|W| = 2n(n-1) \cdot (2^{n-3}(n-2)! \cdot 2) = 2^{n-1}n!.$$

E_6. There are 72 roots in Φ, all of them long; the isotropy group W' is of type A_5. Therefore

$$|W| = 72 \cdot (5+1)! = 72 \cdot 6! = 2^7 \cdot 3^4 \cdot 5.$$

E_7. There are 126 roots in Φ, all of them long; the isotropy group W' is of type D_6 and has order $2^5 \cdot 6!$. Therefore

$$|W| = 126 \cdot 2^5 \cdot 6! = 2^{10} \cdot 3^4 \cdot 5 \cdot 7.$$

E_8. There are 240 roots in Φ, all of them long; the isotropy group W' is of type E_7 and has order $2^{10} \cdot 3^4 \cdot 5 \cdot 7$ (just computed). Therefore

$$|W| = 240 \cdot 2^{10} \cdot 3^4 \cdot 5 \cdot 7 = 2^{14} \cdot 3^5 \cdot 5^2 \cdot 7.$$

F_4. There are 24 long roots; the isotropy group W' is of type C_3 and has order $2^3 \cdot 3!$. Therefore

$$|W| = 24 \cdot 2^3 \cdot 3! = 2^7 \cdot 3^2.$$

$\square$

Exercises

18.1. Prove that the roots in the root systems H_3 and H_4 form a single orbit under the action of the corresponding reflection groups.

18.2. Lemma 18.1 is not true when the root system Φ is not indecomposable. Give an example.

18.3. Give an example of a root system of type $A_1 \oplus A_1 \oplus A_1$ with roots of three different lengths.

18.4. For root systems B_n, C_n, and F_4, find the number of short roots.

18.5. Using the information about the total number of roots in the standard root systems, find, for every irreducible finite reflection group W, the length of the longest element in the group.

Part V

Three-Dimensional Reflection Groups

19

Reflection Groups in Three Dimensions

We give a quick and easy classification of finite systems of mirrors in three dimensions. Our approach is based on some elementary spherical geometry.

This chapter can be read independently from the rest of the book if you start with Chapters 5, 6, and 7 and then jump directly here.

19.1 Planar Mirror Systems

We start with the observation that the classification of planar mirror systems is self-evident. (Compare with Section 8.3.)

Theorem 19.1. *Every finite closed system of mirrors in $\mathbb{R}^2$ is the mirror system of a regular polygon.*

In particular, by Theorem 7.2, the corresponding reflection group is the dihedral group Dih_{2n}.

19.2 From Mirror Systems to Tessellations of the Sphere

Let Σ be a finite closed system of mirrors in the three-dimensional Euclidean space $\mathbb{R}^3$, and let W be the corresponding reflection group. We know that all mirrors in Σ intersect in a common point o. We consider the sphere S of radius 1 centered at o and trace on S the intersections of S with mirrors in Σ; we obtain a *tessellation* of S by spherical polygons, something similar to Figure 19.1.

A few words about spherical geometry are necessary. It is a sister theory to Euclidean geometry, and, because of applications in astronomy, has a long and proud history.

The existence of *antipodal points* is one of the crucial differences from Euclidean plane geometry. Recall that two points a and a' on the sphere are called antipodal to each other if the straight line aa' passes through the center of the sphere, or, in an equivalent form, if the segment $[a, a']$ is a *diameter* of the sphere.

A.V. Borovik and A. Borovik, *Mirrors and Reflections: The Geometry of Finite Reflection Groups*, Universitext, DOI 10.1007/978-0-387-79066-4_19,

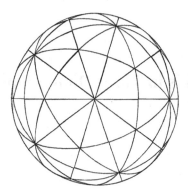

Fig. 19.1. A tessellation of the sphere made from the system of mirrors of the icosahedron.

Great circles, that is, circles cut in the sphere by planes that pass through the center of the sphere, play in spherical geometry a role similar to that of straight lines in Euclidean plane geometry. In particular, if two points a and b are not antipodal, then there is only one great circle that passes through them; this is, of course, the trace in the sphere of the (unique) plane through a, b and the center o of the sphere. On the surface of the sphere, the smaller of the two arcs of the great circle is the shortest path from a to b; but we will not use this property.

For our purpose, of great importance is the *angle* between two great circles, which is, by definition, the angle between the planes containing the circles.

The triangle is the simplest geometric figure in the Euclidean plane; on the sphere, this role is passed to the *digon*, the part of the sphere bounded by two great circles; obviously, the vertices of a digon are antipodal to each other, and the size and shape of the digon is described by just one parameter, its angle; see Figure 19.2.

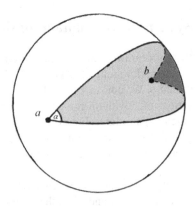

Fig. 19.2. Digon.

Lemma 19.2. *On the sphere of radius* 1, *the area of the digon with angle α is* 2α.

Proof. We start the proof by recalling that the area of the sphere of radius 1 is 4π; the full angle around the point is 2π, and therefore the area of the digon formed by two arcs of great circles with angle α between them is

$$\frac{\alpha}{2\pi} \cdot 4\pi = 2\alpha.$$

$\square$

19.3 The Area of a Spherical Triangle

Theorem 19.3. *On a sphere of radius* 1, *the area of a spherical triangle with angles* α, β, γ *is*

$$\alpha + \beta + \gamma - \pi.$$

Proof. Let α, β, γ be the angles of the triangle T. Let H be the hemisphere containing the triangle determined by the great circle through the vertices of the triangle with angles β and γ (Figure 19.3).

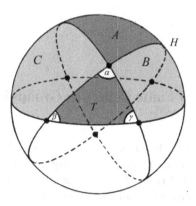

Fig. 19.3. The area of a spherical triangle, Theorem 19.3.

The great circles that continue the sides of the triangle T cut the hemisphere H into our triangle T and spherical polygons A, B, C so that $B \cup T$ and $C \cup T$ are digons of angles β and γ, while A and the triangle T' antipodal to T form the digon of angle α. By Lemma 19.2, the areas $\sigma(B \cup T)$, $\sigma(C \cup T)$ and $\sigma(A \cup T') = \sigma(A \cup T)$ of the three digons are 2β, 2γ, and 2α, respectively. Since the area of the hemisphere is 2π, we have

$$\begin{aligned} 2\pi &= \sigma(T) + \sigma(A) + \sigma(B) + \sigma(C) \\ &= \sigma(T) + (2\alpha - \sigma(T)) + (2\beta - \sigma(T)) + (2\gamma - \sigma(T)) \\ &= 2(\alpha + \beta + \gamma) - 2\,\sigma(T), \end{aligned}$$

whence the result.   □

Corollary 19.4. *The area of a spherical n-gon N is expressed as*

$$\sigma\,(N) = \sum(\text{angles of } N) - (n-2)\pi.$$

Proof. For digons and triangles ($n = 2$ or 3) the statement of the corollary follows from Lemma 19.2 and Theorem 19.3. If $n \geqslant 4$, cut the polygon into triangles and apply Theorem 19.3.   □

Corollary 19.5. *If all angles of a spherical n-gon N do not exceed $\pi/2$ then $n \leqslant 3$.*

Proof. We apply Corollary 19.4:

$$\sigma\,(N) = \sum(\text{angles of } N) - (n-2)\pi \leqslant n \cdot \frac{\pi}{2} - (n-2)\pi$$

has to be positive; hence

$$\frac{n}{2} - (n-2) > 0$$

and $n < 4$.   □

19.4 Classification of Finite Reflection Groups in Three Dimensions

Assume that we have a finite closed mirror system Σ in the 3-dimensional Euclidean space $\mathbb{R}^3$, with all mirrors passing through the origin o; we take intersections of the mirrors with the sphere S of radius 1 centered at o. Have a look at Figure 19.1; the key idea of the classification is to look at possible shapes of spherical polygons cut in the sphere by the great circles (mirrors). We shall slightly abuse the terminology developed in this book and call these polygons *chambers*. Two chambers are *adjacent* if they share a common edge. Obviously, adjacent chambers are mirror images of each other, and as a result, all chambers have the same shape.

Notice further that all great circles meeting at the same vertex are traces on S of mirrors forming a mirror system on its own, say Σ'. All mirrors in Σ' intersect in a common straight line l; looking at the intersections of mirrors with a plane P perpendicular to l, we get a closed mirror system Σ'' in P. By Lemma 8.4, the angles between neighboring mirrors in Σ'', hence in Σ', are all equal and therefore do not exceed $\pi/2$. This means, of course, that chambers are digons or triangles; see Corollary 19.5.

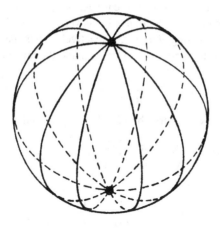

Fig. 19.4. *Mirror system of type $G_2(m)$.*

If the chambers are digons, then they all share two vertices, which can be conveniently called the north and the south pole. In that case, all mirrors in Σ contain the axis through the poles, and Σ is the mirror system of a dihedral group (Figure 19.4). If Σ contains m mirrors, its reflection group is the dihedral group Dih_{2m} of order $2m$, and the traditional notation for the mirror system is $I_2(m)$ or $G_2(m)$ (in this book, we use the latter).

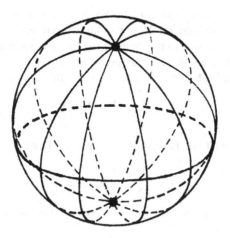

Fig. 19.5. *Mirror system of type $G_2(m) + A_1$.*

The principal and most interesting case is, of course, that of chambers being triangles. Fix some chamber C; let a, b, c be its vertices, and denote by m_1, m_2, m_3 the number of mirrors meeting at a, b, c, respectively. Then the angles at the vertices are equal, respectively, to π/m_1, π/m_2, π/m_3, and the area of the triangle is given by

$$\sigma(C) = \frac{\pi}{m_1} + \frac{\pi}{m_2} + \frac{\pi}{m_3} - \pi.$$

Let N be the number of chambers; then $S(C) = 4\pi/N$ and we arrive at the equation

$$\frac{4\pi}{N} = \frac{\pi}{m_1} + \frac{\pi}{m_2} + \frac{\pi}{m_3} - \pi,$$

which simplifies to

$$1 + \frac{4}{N} = \frac{1}{m_1} + \frac{1}{m_2} + \frac{1}{m_3}.$$

In particular,

$$\frac{1}{m_1} + \frac{1}{m_2} + \frac{1}{m_3} > 1;$$

this inequality has only finitely many solutions in positive integers $m_1 \geqslant 2$, $m_2 \geqslant 2$, $m_3 \geqslant 2$, and all solutions can be easily listed by direct inspection. Notice that every triple (m_1, m_2, m_3) immediately yields the corresponding value of N. We have freedom of notation and can assume that $m_1 \geqslant m_2 \geqslant m_3$. After that, the list of all solutions becomes very compact. The first solution is

$$1 + \frac{4}{4m} = \frac{1}{2} + \frac{1}{2} + \frac{1}{m}. \tag{19.1}$$

The corresponding mirror system is $G_2(m)$ with the added "equatorial" mirror; see Figure 19.5. The corresponding reflection group is the direct product of the dihedral group Dih_{2m} (it corresponds to the "meridional" mirrors) and the group of order 2 (generated by the reflection in the "equatorial" mirror).

The next solution leads to the system of mirrors of the regular tetrahedron:

$$1 + \frac{4}{24} = \frac{1}{2} + \frac{1}{3} + \frac{1}{3}. \tag{19.2}$$

Moving further, we have the system of mirrors of the cube:

$$1 + \frac{4}{48} = \frac{1}{2} + \frac{1}{3} + \frac{1}{4}. \tag{19.3}$$

Our final solution leads to the mirror system presented in Figure 19.1:

$$1 + \frac{4}{120} = \frac{1}{2} + \frac{1}{3} + \frac{1}{5}. \tag{19.4}$$

This is the mirror system of the *icosahedron*; we discuss its construction in Chapter 20.

Exercises

19.1. Mark takes under m hours to do a job, Nick takes under n hours for the same job, and Len takes under l hours for the job. If m, n, l are integers, what are the possible values of m, n, and l that ensure that Len, Mark, and Nick, working together, complete the job in well under one hour?

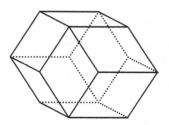

Fig. 19.6. A rhombic polyhedron.

19.2. Find the group of symmetries of the rhombic polyhedron in Figure 19.6.

19.3. Find all (convex) polytopes Δ in $\mathbb{R}^3$ whose vertices can be cyclically permuted, that is, for which there is a cyclic group C in $\mathrm{Sym}(\Delta)$ such that C acts transitively on the set of vertices of Δ.

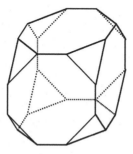

Fig. 19.7. A truncated cube is an example of a vertex-transitive polytope.

Projects

Here, we have collected a few more serious problems; some of them are much harder than they appear. Their solutions require a systematic approach; we recommend using the problems as long term group projects.

19.4. A *half-turn* is a rotation through $180°$ about an axis; obviously, half-turns can be alternatively characterized as orthogonal transformations with eigenvalues $1, -1, -1$. Classify the finite groups of orthogonal transformations of $\mathbb{R}^3$ that are generated by half-turns.

19.5. Classify and sketch convex vertex-transitive polytopes in $\mathbb{R}^3$ (Figure 19.7).

19.6. Classify and sketch edge-transitive polytopes in $\mathbb{R}^3$.

19.7. Classify and sketch face-transitive polytopes in $\mathbb{R}^3$.

20

Icosahedron

In this chapter we give a construction of the icosahedron that is both self-evident and rigorous. This is a surprisingly neglected topic. Pictures of icosahedra are abundant in books on geometry, and they create in the reader the false impression that a beautiful picture proves the existence of the object.

Moreover, we separate clearly the issues of *existence* of the icosahedron and its *uniqueness*. As we shall see, the uniqueness is intimately related to the fact that the icosahedron has a rich set of symmetries that only a polytope can admit.

20.1 Construction

For the construction of an icosahedron, we follow, with some modifications, the method of H. M. Taylor[1] [Hea, pp. 491–492].

The attractive feature of Taylor's method is that it gives the most effective way of drawing an icosahedron, so simple that it is accessible to the reader with very modest drawing skills. As shown in Figure 20.1, first we mark symmetrically positioned segments in an alternating fashion on the faces of the cube (left), and then connect the endpoints (right).[2]

The drawing actually provides a proof of the existence of the icosahedron: varying the lengths of segments on the left cube, it is easy to see from continuity principles (Figure 20.2) that at a certain length of the segments, all edges of the inscribed polytope on the right become equal. Therefore we get a convex polytope Δ with the following properties:

(a) all faces of Δ are equilateral triangles, and
(b) five faces meet at each vertex.

---

[1] A referee kindly commented that this method actually goes back to Piero della Francesca (1416–1492).

[2] In this chapter, we reserve the term "face" for 2-dimensional faces of polytopes, deviating from the general terminology of Section 3.1.

A.V. Borovik and A. Borovik, *Mirrors and Reflections: The Geometry of Finite Reflection Groups*, Universitext, DOI 10.1007/978-0-387-79066-4_20,

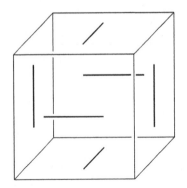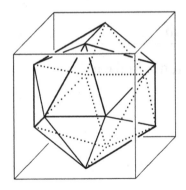

Fig. 20.1. A self-evident construction of an icosahedron

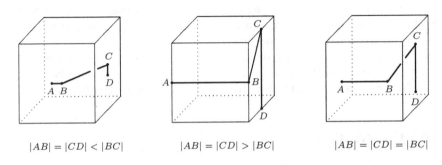

$|AB| = |CD| < |BC|$ \qquad $|AB| = |CD| > |BC|$ \qquad $|AB| = |CD| = |BC|$

Fig. 20.2. The continuity principle says that the distance between two points is a continuous function of the coordinates of the points and therefore assumes all intermediate values (compare with Figure 20.3).

This is what is called an *icosahedron*. However, the difficulties start as soon as we think about what we have constructed.

> *Is the icosahedron* unique? *More precisely, if we fix the length of the edge, do properties* (a) *and* (b) *define the icosahedron uniquely up to isometry?*

For example, why does Taylor's method produce the same result as another well-known method, due to Kepler (see Figure 20.4)?

Until we have clarified this issue, we shall call the icosahedron obtained by inscription in the cube *Taylor's polytope* and denote it by Δ; similarly, *Kepler's polytope* will be denoted by Γ.

What are the groups of symmetries of Δ and Γ?

It is easy to see that $\mathrm{Sym}(\Delta)$ acts transitively on the set of vertices of Δ; moreover, the group of symmetries of the ambient cube contains a subgroup of order 24 that preserves Δ and acts transitively on the set of vertices of Δ.

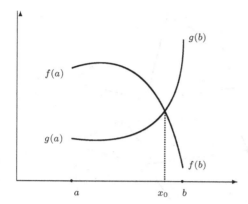

Fig. 20.3. The intermediate value theorem: if f and g are continuous functions on $[a, b]$ and $f(a) > g(a)$, $f(b) < g(b)$, then there exists a point $x_0 \in [a, b]$ such that $f(x_0) = g(x_0)$.

But is $\mathrm{Sym}(\Delta)$ *transitive on the set of edges? On the set of faces?*

Similarly, we see that in Kepler's construction, $\mathrm{Sym}(\Gamma)$ admits a rotation of order 5 about the north pole – south pole axis.

But is $\mathrm{Sym}(\Gamma)$ *transitive on the set of vertices? On the set of edges? On the set of faces?*

20.2 Uniqueness and Rigidity

We have already stated, without proof, some important and self-evident geometric theorems that are, however, very hard to prove; see, for example, Theorem 3.9. Here is another classical example, due to the famous French mathematician Cauchy. When you make a convex polytope out of cardboard, one of the useful (or annoying, depending on circumstances) properties of an unfinished model is that a polyhedral cone formed by four or more faces sharing a common vertex is *flexible* around the edges; however, the completed cardboard model is *rigid*. This observation can be expressed by saying that a convex polytope is uniquely determined, up to isometry, by its faces and how they are joined together. An accurate mathematical statement is a bit more technical.

Theorem 20.1. (Cauchy's rigidity theorem, [Ber, Theorem 12.8.1]) *Let Δ and Δ' be two convex polytopes and $\mathcal{F}, \mathcal{F}'$ the sets of their faces. Assume that there is a map $\alpha : \mathcal{F} \longrightarrow \mathcal{F}'$ such that*

- α *takes vertices into vertices, edges into edges, and faces into faces;*

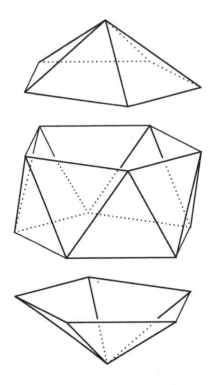

Fig. 20.4. Kepler's construction of the icosahedron: you first make a pentagonal antiprism with all edges equal—it obviously exists by the continuity principle, compare with Figures 20.1 and 20.2—and add two pentagonal pyramids, again with all edges equal. Why does Kepler's method produce the same result as Taylor's?

- *α preserves the adjacency of faces: the common edge of two neighboring faces is mapped to the common edge of the images of the faces;*
- *finally, for every face F of Δ, there is an isometry $\iota_F : F \longrightarrow \alpha(F)$ which agrees with the map α on all edges of F: if E is an edge then the image $\iota_F(E)$ of E coincides with $\alpha(E)$.*

Then there is an isometry $\iota : \Delta \longrightarrow \Delta'$ that agrees with α, that is, if F is an arbitrary face of Δ then the image $\iota(F)$ of F coincides with $\alpha(F)$.

If you accept Cauchy's theorem, then it should be obvious that if Taylor's polytope and Kepler's polytope have equal edge lengths then the polytopes are isometric, because the way they are assembled from equilateral triangles is the same for both polytopes.

We can claim even more: any two convex polytopes with 20 equilateral triangular faces and 12 vertices arranged in such way that exactly 5 facts meet at each vertex, are isometric if they have equal edge lengths. Therefore we indeed have the right to apply the term *icosahedron* to all such polytopes.

In a twist typical for mathematics, we can apply Cauchy's theorem to a single icosahedron Δ: if $[V, E, F]$ is a *flag of faces* in Δ, that is, the vertex V is an endpoint of the edge E that is a side of the face F, and $[V', E', F']$ is another flag, then it is intuitively clear that there is a map α that satisfies the conditions of Cauchy's theorem and sends V to V', E to E', and F to F'. Moreover, it is easy to see that such a map is unique. Therefore, by Cauchy's theorem, there is a *unique* isometry of Δ that sends the flag $[V, E, F]$ to $[V', E', F']$. In group-theoretic terminology, Sym Δ acts simply transitively on the set of flags and hence the order of Sym Δ equals the number of flags. The latter, obviously, is 12 (the number of vertices) times 5 (the number of edges coming out of the vertex) times 2 (the choice of two faces meeting at the edge), which gives us 120.

Furthermore, Cauchy's theorem allows us to check that Sym Δ is generated by reflections—we leave checking the details to the reader.

20.3 The Symmetry Group of the Icosahedron

In this section we will show, by a direct calculation, that Taylor's construction and Kepler's construction lead to the same result. To that end, we compute the length of the edge of the icosahedron inscribed in the cube with edge of length 2.

It will convenient for us to work with the cube $[-1, 1]^3$ formed by the planes $x = \pm 1$, $y = \pm 1$, $z = \pm 1$; in Figure 20.5, the cube is turned to the viewer showing its faces $x = 1$, $y = 1$, $z = 1$.

If now we mark symmetrically positioned segments of length 2α on the faces of the cube, then the coordinates of their points will be as shown in Figure 20.5. To ensure that all edges of the inscribed polytope have equal length, it will suffice to check that $|AB'| = |BB'|$, or in coordinate form,

$$\sqrt{(1 - 0)^2 + (\alpha - 1)^2 + (0 + \alpha)^2} = 2\alpha,$$

which after simplification gives us

$$\alpha^2 + \alpha - 1 = 0,$$

the latter equation having only one positive root,

$$\alpha = \frac{-1 + \sqrt{5}}{2}.$$

Now a miracle happens: if O is the origin of the coordinate system, then we see that the scalar products

$$\overrightarrow{OC'} \cdot \overrightarrow{OB} = \overrightarrow{OC} \cdot \overrightarrow{OB} = \overrightarrow{OA} \cdot \overrightarrow{OB} = \overrightarrow{OB'} \cdot \overrightarrow{OB} = \alpha$$

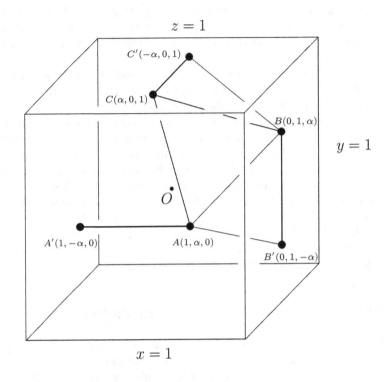

Fig. 20.5. Calculating the icosahedron inscribed into the cube $[-1,1]^3$: if the length of the segments marked on the faces of the cube is 2α then the coordinates of their endpoints on the faces $x = 1$, $y = 1$, $z = 1$ are as shown in the diagram.

are all equal. Hence the points C', C, A, B' all belong to the same plane perpendicular to the vector $\overrightarrow{OB'}$. The fifth vertex of the inscribed polytope adjacent to B (it is not shown in the diagram) has coordinates $(-1, \alpha, 0)$ and obviously belongs to the same plane.

Now it is obvious that B and five adjacent vertices of Taylor's polytope form a pentagonal pyramid with equal edges; if we remove from Taylor's polytope vertex B and its opposite, we get Kepler's pentagonal antiprism; see Figure 20.6.

Hence Taylor's polytope is also Kepler's! Even if we are not using Cauchy's rigidity theorem, we have right to call our polytope an *icosahedron*.

Now we have no difficulty in determining the symmetry group of the icosahedron Δ. We have already observed that Taylor's construction ensures that $\mathrm{Sym}(\Delta)$ contains a subgroup of order 24 that acts transitively on the set of vertices of Δ. We have also observed that Kepler's construction obviously has rotational symmetry: five faces of the icosahedron meeting at its "north pole" (Figure 20.4) can be moved one to the other by a rotation around the pole. Combining these two observations, we see

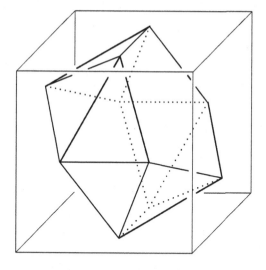

Fig. 20.6. Pentagonal antiprism inscribed in the cube.

that Sym Δ acts transitively not only on vertices, but also on edges and faces of the icosahedron.

The concept of a flag (vertex–edge–face) introduced in the previous section is again quite useful. Indeed, combining the symmetries we get from Taylor's and from Kepler's constructions, we can further observe that Sym Δ acts simply transitively on the set of flags of the icosahedron and therefore has order 120.

Exercises

20.1 (Armstrong). Glue two dodecahedra together along a pentagonal face and find the rotational symmetry group of this new solid. What is its full symmetry group?

Part VI

Appendices

Part VI

A

The Forgotten Art of Blackboard Drawing

As the reader has possibly noticed, numerous drawings in this book are very simple, almost primitive. This is a conscious choice on the part of the authors; indeed, what matters—and it is part of our teaching philosophy—is that the pictures are *reproducible*.

We probably have to emphasize the difference between *drawings* or *sketches*, which are supposed to be reproduced by the reader or student, and more technically sophisticated illustrative material (which we shall call *illustrations*), especially computer-generated images designed for visualization of complex mathematical objects. It would be foolish to impose restrictions on the technical perfection of illustrations. However, we believe that *drawings* should be intentionally made very simple, almost primitive. They should not instill an inferiority complex in the reader who has not attempted to draw anything since the halcyon days of elementary school. Even if the reader has very modest drawing skills, he or she should be able to draw similar pictures as a way of facilitating study.

In this appendix, we give some advice on making usable mathematical drawings. First of all,

Treat your sketch as a mathematical object.

Indeed, most pictures in this book are parallel projections of three-dimensional polytopes on a plane; if you specify the polytope, the plane, and the direction of projection, then the projection is a well-defined mathematical figure.

However, we have to take into account some peculiarities of human visual perception, as well as cultural traditions. Therefore our next advice is very general and entirely nonmathematical.

If you a right-hander, then in your drawings, show a three-dimensional body as if you are holding it in your left hand: slightly lower than your viewpoint and slightly to the left.

The rationale behind this advice is obvious: this is how right-handed people see an apple when they pare it with a knife held, naturally, in the right hand.

A.V. Borovik and A. Borovik, *Mirrors and Reflections: The Geometry of Finite Reflection Groups*, Universitext, DOI 10.1007/978-0-387-79066-4,

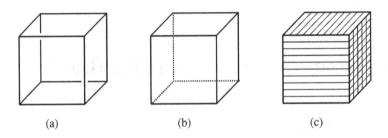

Fig. A.1. Various ways to emphasise the depth of a picture: (a) wireframe model, (b) semitransparent filling, (c) shading.

We do not know what say to left-handed readers; like most things in this world, the pictures in this book are designed for right-handers. Perhaps you should simply adopt our advice with the obvious change of left and right. Of course, a lot depends on whether you draw pictures for your own personal use, or, say, for your students. Also, even if you are left-handed, it might be that you have already been conditioned by the thousands of images you have seen in books and magazines, and are more comfortable when looking at the world from a right-hander's point of view.

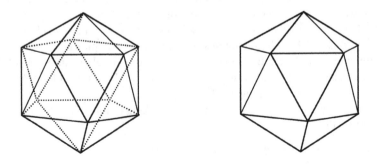

Fig. A.2. As a sketch on a blackboard, a semitransparent polytope (left) is more convincing than a filled one (right).

Emphasize the depth of a picture by some simple method (Figure A.1):
- *Wireframe model: polytopes (or, more precisely, their edges) made of wire, with breaks showing where one piece of wire passes behind another.*
- *Semitransparent filling: edges obscured by faces of the polytope shown by dotted lines; see Figure A.2.*
- *Shading. Illustrators usually recommend placing the brightest spot on the front of the object.*

Notice that the shading of faces of the *stella octangula* in Figure A.3 does not follow the last rule: we sacrificed it for the sake of simplicity of the picture.

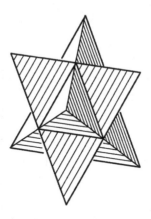

Fig. A.3. Shading the faces of stella octangula.

Inscribe more complicated polytopes within simpler ones; again, mathematical relations between various polytopes could be very helpful for understanding more complex polytopes.

Taylor's construction of the icosahedron is our favorite application of this principle; see Figure 20.1. Another illustration is a construction of the *stella octangula*, Figure A.4.

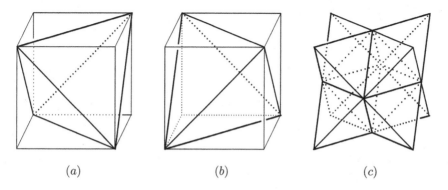

(a) (b) (c)

Fig. A.4. Drawing stella octangula.

As you can see, there is nothing really difficult here; just dare to believe that you can draw.

Exercises

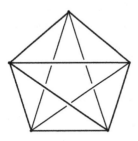
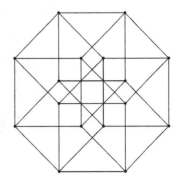

Fig. A.5. For Exercise A.1: make a wireframe model of a 4-dimensional cube.

A.1. In Figure A.5, the picture on the left is a wireframe model of a 4-dimensional simplex; notice that the line breaks, which indicate which edge of the simplex is farther away from the 2-dimensional plane of projection. Which lines have to be broken in the picture on the right in order to make it a realistic wireframe model of a 4-dimensional cube?

B

Hints and Solutions to Selected Exercises

1.1 *Hint:* Consider $\mathbb{C}^2$ as a vector space V over $\mathbb{R}$. Then it is a 4-dimensional vector space over $\mathbb{R}$, and a 1-dimensional subspace U over $\mathbb{C}$ becomes a 2-dimensional subspace over $\mathbb{R}$. Choose a basis $\epsilon_1, \ldots, \epsilon_4$ in V in such a way that ϵ_1, ϵ_2 is a basis of U. If now $\alpha = (a_1, a_2, a_3, a_4)^T$ and $\beta = (b_1, b_2, b_3, b_4)^T$ are two points in $V \smallsetminus U$, then at least one of the coordinates a_1, a_2 and at least one of b_1, b_2 are different from 0. What remains is to find a path connecting α and β; for example, take a point $\gamma = (c_1, c_2, c_3, c_4)$ sufficiently far away from U (which simply means that one of $|c_3|, |c_4|$ is sufficiently big). Then such a path is formed by two segments $[\alpha, \gamma] \cup [\gamma, \beta]$.

1.3 Solution: We can assume without loss of generality that the origin belongs to A $o \in A$. If A is an affine subspace, then A is a vector subspace of $\mathbb{R}^n$, and in that case, of course, for any two distinct points $a, b \in A$, the line $a + \mathbb{R}\overrightarrow{ab}$ through a and b belongs to A. On the other hand, if α, β are vectors in A with endpoints a and b, then every vector $k\alpha$ for $k \in \mathbb{R}$ belongs to a line trough o and the endpoint a of α. Furthermore, A is closed under taking sums of vectors for the following reason:

$$\alpha + \beta = 2 \cdot \left(\frac{1}{2}(\alpha + \beta) \right),$$

and the endpoint of $\frac{1}{2}(\alpha + \beta)$ is the midpoint of segment $[a, b]$ and therefore belongs to A.

1.4 Solution: Let π be an orthogonal projection of $\mathbb{R}^n$ onto A. If $a, b \in \pi[X]$, take their preimages $x, y \in X$. Then a segment $[x, y]$ belongs to X by definition of a convex set. The image $\pi[[x, y]]$ of $[x, y]$ is a segment connecting a and b (prove!); obviously, it belongs to $\pi[X]$. Therefore $\pi[X]$ is connected.

1.5 Solution: Assume that we fix the first mirror to a workbench and polish it with the second one. Then the fixed mirror will tend to take a convex shape, while the moving mirror will develop a complimentary concave shape. Rotating three mirrors in this process, we change their roles and make them both concave and convex, that is, flat.

2.1 *Hint:* Drop a perpendicular from the vertex c to the side $[a, b]$.

2.2 *Hint:* Introduce orthonormal coordinates $x_1, \ldots, x_n$ and show that the system of equations

$$\frac{\partial M(x)}{\partial x_i} = 0, \quad i = 1, \ldots, n,$$

is equivalent to the equation

A.V. Borovik and A. Borovik, *Mirrors and Reflections: The Geometry of Finite Reflection Groups*, Universitext, DOI 10.1007/978-0-387-79066-4,
© Springer Science+Business Media, LLC 2010

$$\sum_{j=1}^{k} x\overrightarrow{f_j} = 0,$$

where $x = (x_1, \ldots, x_k)$.

2.7 *Hint:* This is an immediate consequence of Theorem 2.3.

2.11 *Hint:* If an orthogonal transformation of $\mathbb{R}^3$ has determinant $+1$ then it has eigenvalues

$$+1, \quad \cos\theta + i\sin\theta, \quad \cos\theta - i\sin\theta,$$

and therefore is a rotation through the angle θ about the 1-dimensional eigenspace for eigenvalue $+1$.

3.1 *Hint:* Use induction on n.

3.2 *Hint:* Use induction on the number n of lines. The basis of induction ($n = 1$) is obvious. After the next line is drawn, change the color of all chambers on one side of it.

3.4 *Hint:* An immediate application of the definition of the gallery distance.

3.5 *Hint:* Use Proposition 3.4 and Lemma 3.3.

3.6 *Hint:* When answering the second question, consider first the 2-dimensional case, Figure 3.1.

Answer: Let C be a chamber in Σ. Then $\overline{C} \cap \Delta$ is the closure of a face of Δ, and every chamber in Σ is uniquely defined by the (non-empty!) set of vertices of Δ that belong to $\overline{C}$. There are $2^4 - 1 = 15$ nonempty subsets of the set of four points.

3.7 *Hint:* For a point $x = (x_1, \ldots, x_{n+1})$ in A that does not belong to any of the hyperplanes $x_i = 0$, look at all possible combinations of the signs $+$ and $-$ of the coordinates x_i of x, $i = 1, \ldots, n+1$.

3.8 Answers: 1. For example, a rectangular box with nonsquare faces.

2. Glue two congruent regular tetrahedra over a common face.

3. For example, the cuboctahedron of Figure 9.9.

3.9 Solution: If a symmetry of a polytope Δ fixes every vertex of Δ then it fixes every point of Δ (because Δ is the convex hull of its vertices). Therefore nonidentity symmetries of Δ act as nonidentity permutations of vertices of Δ, and different symmetries act as different permutations (why?). But there are at most $n!$ permutations of n points.

3.10 Answer (one of many possible): In $A\mathbb{R}^2$, with Cartesian coordinates (x, y), the stripe

$$\{ (x, y) : 0 \le x \le 1 \}$$

is the intersection of two closed half-spaces

$$\{ (x, y) : 0 \le x \}$$

and

$$\{ (x, y) : x \le 1 \}$$

and therefore is a polyhedron. It is invariant under (infinitely many) translations along the x-axis.

5.6 *Hint:* It is generated by a rotation through the smallest angle.

5.7 Solution: By Exercises 5.2 and 5.3,

$$\det sr = \det s \cdot \det r = -1;$$

hence sr is a reflection and $(sr)^2 = 1$. But then $srs \cdot r = 1$ and $r^s \cdot r = 1$. Hence $r^s = r^{-1}$.

5.9 *Hint:* Use Exercise 5.3 and multiply the matrices of the two reflections

$$\begin{pmatrix} \cos\phi & \sin\phi \\ \sin\phi & -\cos\phi \end{pmatrix} \quad \text{and} \quad \begin{pmatrix} \cos\theta & \sin\theta \\ \sin\theta & -\cos\theta \end{pmatrix}.$$

6.1 Answer: It does not change left and right, it changes front and back.

6.2 *Hint:* If M and N are two mirrors in Σ with the reflections s and t, then in view of Lemma 5.3, the mirror image of M in N is the mirror of the reflection s^t. If reflections s and t map Δ onto Δ then so does s^t.

6.5 *Hint:* the case of α/π irrational requires some special care.

6.6 Solution: Each reflection changes the sign of one of the components of the direction vector of the ray of light; after three reflections, the direction changes to exactly the opposite one.

6.7 Answers: 6 and 9.

6.8 Answer: 15. Each mirror of symmetry passes through a unique pair of opposite edges.

6.12 *Hint:* Choose Cartesian coordinates in $\mathbb{AR}^2$ in such a way that parallel mirrors H_1 and H_2 have coordinates $x = 1$ and $x = 2$.
Solution: Then a direct calculation shows that the product $s_1 s_2$ of the corresponding reflections is the translation through a vector perpendicular to H_1 and H_2, directed from H_1 to H_2 and of length equal to twice the distance between H_1 and H_2.

6.13 *Hint:* Check first that in a bounded figure, the center of symmetry lies on a mirror of symmetry. Then select an orthonormal coordinate system with the origin at that center of symmetry, and represent both central symmetry and mirror symmetry by matrices. What are their eigenvalues? What are the eigenvalues of their product?
Hint for $\mathbb{AR}^3$: It is no longer true, but we can claim that the body has an axis of a half-turn, rotation through $180°$.

7.1 *Hint:* Make Sym_3 act on the three vertices of an equilateral triangle (regular 3-gon).

8.1 Solution:

$$s_\alpha\beta \cdot s_\alpha\gamma = \left(\beta - \frac{2\beta \cdot \alpha}{\alpha \cdot \alpha}\alpha\right) \cdot \left(\gamma - \frac{2\gamma \cdot \alpha}{\alpha \cdot \alpha}\alpha\right)$$

$$= \beta \cdot \gamma - \frac{2\gamma \cdot \alpha}{\alpha \cdot \alpha}(\beta \cdot \alpha) - \frac{2\beta \cdot \alpha}{\alpha \cdot \alpha}(\alpha \cdot \gamma) + \frac{2\beta \cdot \alpha}{\alpha \cdot \alpha} \times \frac{2\gamma \cdot \alpha}{\alpha \cdot \alpha}(\alpha \cdot \alpha)$$

$$= \beta \cdot \gamma.$$

8.5 *Hint:* Find a regular n-gon such that $W(\Phi)$ coincides with its symmetry group.

8.6 *Hint:* Use Exercise 7.7.

8.7 *Hint:* Use Exercise 7.7.

8.9 *Hint:* The two groups Dih_6 correspond to the short root subsystem and long root subsystem of the root system G_2, see Figure 8.4.

9.1 *Hint:* Consider a figure formed by neighboring vertices (and edges connecting them) of any of the vertices (Figure B.1).

9.2 Solution: Figures B.2 and B.3.

9.5 *Hint:* Notice that a reflection is an involution; hence every cycle in its cycle decomposition on $[n] \sqcup [n]^*$ is of length 1 or 2.

9.6 Solution: Indeed, it is the group of symmetries of a square (regular 4-gon).

9.15 Answer: BC_3.

10.1 *Hint:* Notice that the cone

$$x_{i_1} < x_{i_2} < \cdots < x_{i_n}$$

is bound by $n - 1$ walls

$$x_{i_1} = x_{i_2}, \quad x_{i_2} = x_{i_3}, \quad \ldots, \quad x_{i_{n-1}} = x_{i_n}$$

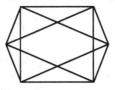

Fig. B.1. 2-dimensional projection of a 3-dimensional cross polytope.

Fig. B.2. Root systems $A_1 \oplus A_1$.

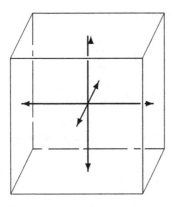

Fig. B.3. The root system $A_1 \oplus A_1 \oplus A_1$ inscribed in the unit cube $[-1, 1]^3$.

and can be described as

$$C = \{ \gamma \mid \gamma \cdot (\epsilon_{i_j} - \epsilon_{i_{j+1}}), \; j = 1, 2, \ldots, n-1 \},$$

where

$$\{ \epsilon_{i_j} - \epsilon_{i_{j+1}}, \; j = 1, 2, \ldots, n-1 \}$$

is a simple system of vectors.

11.1 *Hint:* We already know that the roots

$$\epsilon_2 - \epsilon_1, \epsilon_3 - \epsilon_2, \ldots, \epsilon_n - \epsilon_{n-1}$$

form a simple system of roots in the root system A_{n-1}.

11.2 *Hint:* Similarly to the previous exercise, check that these reflections correspond to roots in the simple root system of type BC_n.

11.3 *Hint:* Similar to the two previous exercises.

11.8 *Hint:* Use induction on l.

12.2 Solution: Indeed this is a permutation with the maximal possible number of inversions.

12.3 *Hint:* We know that w_0 sends the fundamental chamber to the opposite chamber, $w_0 C = -C$. Where does w_0 send $-C$?

12.8 Answers: Dihedral groups of orders 6, 4, 8 respectively.

12.13 *Hint:* Decompose the space V into the direct sum of eigenspaces for t: $V = V_{+1} \oplus V_{-1}$. Observe that $t \in C_W(V_{+1})$ and that this group is generated by reflections by Theorem 12.6. Observe further that $C_W(V_{+1})$ leaves the subspace $V_{-1} = V_{+1}^\perp$ invariant, and apply induction on $\dim V$.

13.5 *Hint:* Use Exercise 12.1.

15.1 *Hint:* this is where your paper models come in most handy. Use them to trace consecutive contractions. Make your own examples of similar calculations in D_3.

17.1 *Hint:* Notice that W is generated by simple reflections; therefore in view of Proposition 17.1, it suffices to check (17.1) only for simple roots α, β, hence only for planar root systems of types A_2, B_2, C_2, G_2, and $A_1 \oplus A_1$.

17.4 Answer: By direct comparison of root systems, if Φ is of type B_n, then Φ_{long} is D_n. The system of short roots Φ_{short} in B_n consists of pairwise orthogonal pairs of roots $\pm\epsilon_i$ and is therefore

$$A_1 \oplus \cdots \oplus A_1.$$

In C_n, conversely: Φ_{short} is D_n and Φ_{long} is $A_1 \oplus \cdots \oplus A_1$.
The root system F_4 consists of 24 long roots

$$\pm\epsilon_i \pm \epsilon_j \quad (i < j)$$

forming a system D_4, just by definition of the root system D_4.
Also, 24 short roots

$$\pm\epsilon_i, \quad \frac{1}{2}(\pm\epsilon_1 \pm \epsilon_2 \pm \epsilon_3 \pm \epsilon_4),$$

form a system of type D_4, although it is harder to see. However, observe:
- all roots in Φ_{short} are of the same length;
- dot products of noncollinear roots take values 0 or $\pm 1/2$ (direct computation);
- there are 24 short roots.

A direct comparison with lists of the root systems in Chapter 17 shows that the only possibility is D_4.

17.6 *Hint:* Observe that the set of values of scalar squares $\lambda \cdot \lambda$ for $\lambda \in \Lambda$ is a discrete subset of $\mathbb{R}$; hence the lengths of nonzero vectors in Λ are bounded from below.

17.8 *Hint:* Observe that the corresponding reflection group contains a rotation through the angle $2\pi/m$, which cannot be written by a matrix with integer entries. The latter can be seen from the values of the roots of the characteristic polynomials of integer 2×2 matrices of determinant 1.

18.1 Solution: All edges in the Coxeter diagram have label 3 ("single" edges). The result now follows from Lemma 18.1.

18.2 Solution: See the next exercise.

18.3 Solution: For example, take in $\mathbb{R}^3$ a root system

$$\Phi = \{\,\pm\epsilon_1, \pm 2\epsilon_2, \pm 3\epsilon_3\,\}.$$

18.4 Solution: Short roots in B_n are $\pm\epsilon_i$; therefore there are $2n$ of them. Notice that short roots of C_n are long roots of B_n and therefore have the form $\pm\epsilon_i \pm \epsilon_j$ for $i \neq j$, which makes $2n(n-1)$ roots.

Let $\epsilon_1, \ldots, \epsilon_4$ be the standard basis in $\mathbb{R}^4$. The root system of type F_4i consists of 24 long roots

$$\pm\epsilon_i \pm \epsilon_j \quad (i < j)$$

and 24 short roots

$$\pm\epsilon_i, \quad \frac{1}{2}(\pm\epsilon_1 \pm \epsilon_2 \pm \epsilon_3 \pm \epsilon_4).$$

18.5 *Hint:* The length sought equals the number of walls separating the fundamental chamber from its opposite, that is, the total number of mirrors.

Solution: The longest element in W corresponds to the opposite chamber $-C$ in the chamber system of W. The opposite chamber $-C$ is separated from the fundamental chamber C by every wall in the mirror system Σ of W. The length of the longest element is the number of walls intersected by a geodesic gallery Γ from the fundamental chamber to the opposite chamber. Since a geodesic gallery intersects every wall only once, Γ intersects, and only once, every mirror in Σ. Hence its length is the number of mirrors in Σ, that is, half the number of roots in the mirror system Φ of W.

References

Coxeter Groups and Reflection Groups

[Bou]   N. Bourbaki, *Groupes et Algebras de Lie, Chap. 4, 5, 6*. Hermann, Paris, 1968.
[Dol]   I. V. Dolgachev, Reflection groups in algebraic geometry, Bull. Amer. Math. Soc., 45 no. 1 (2008) 1–60.
[GB]   L. C. Grove and C. T. Benson, *Finite Reflection Groups*. Springer-Verlag, 1984.
[Hum] J. E. Humphreys, *Reflection Groups and Coxeter Groups*. Cambridge University Press, 1990.
[Kane] R. Kane, *Reflection Groups and Invariant Theory*. Springer, 2001.
[Vin]   E. B. Vinberg, Kaleidoskopy i gruppy otrazhenii, *Matematicheskoe Prosveshchenie*, 3 no. 7 (2003) 45–63 (in Russian).

Group Theory and Geometry

[Arm]   M. A. Armstrong, *Groups and Symmetry*. Springer-Verlag, 1988.
[BGW] A. V. Borovik, I. M. Gelfand and N. White, *Coxeter Matroids*, Birkhäuser, 2003.
[Ber]   M. Berger, *Géométrie*. Nathan, 1990.
[Cox]   H. S. M. Coxeter, *Regular Polytopes*. Methuen and Co., London, 1948.
[Cro]   P. R. Cromwell, *Polyhedra*. Cambridge University Press, 1997.
[Hea]   T. L. Heath, *The Thirteen Books of Euclid's Elements, with Introduction and Commentary*. Cambridge: at the University Press, 1908.
[HT]   D. W. Henderson and D. Taimina, *Experiencing Geometry: Euclidean and Non-Euclidean with History*, 3rd ed., Prentice-Hall, 2004.
[Roc]   R. T. Rockafellar, *Convex Analysis*. Princeton Univerity Press, 1970.
[Tits] J. Tits, A local approach to buildings, in *The Geometric Vein (Coxeter Festschrift)*. Springer-Verlag, New York, 1981, 317–322.

Index